Terminology on
Agricultural Meteorology and Agronomy

Terminology on Agricultural Meteorology and Agronomy

V. Radha Krishna Murthy Ph.D.
Associate Professor
Department of Agronomy, College of Agriculture,
ANGRAU, Rajendranagar, Hyderabad - 500 030

M. Yakadri Ph.D.
Associate Professor
Department of Agronomy, College of Agriculture,
ANGRAU, Rajendranagar, Hyderabad - 500 030

P.V.V. Prasad Ph.D.
Asst. Professor
Kansas State University, Kansas, USA.

BSP **BS Publications**
A unit of **BSP Books Pvt., Ltd.**

4-4-309/316, Giriraj Lane, Sultan Bazar,
Hyderabad - 500 095
Phone : 040 - 23445605, 23445688

© 2015, *by Publisher*

Published by :

BSP **BS Publications**
A unit of **BSP Books Pvt., Ltd.**

4-4-309/316, Giriraj Lane, Sultan Bazar,
Hyderabad - 500 095
Phone : 040 - 23445605, 23445688
e-mail : info@bspbooks.net

ISBN : 978-93-52300-35-8 (HB)

ACHARYA N.G. RANGA AGRICULTURAL UNIVERSITY

Dr. S. RAGHU VARDHAN REDDY
VICE-CHANCELLOR

Foreword

Farmers across the globe, India and in Andhra Pradesh are expected to manage the more insidious effects of long-term climate change and variability that may now be occurring at an unprecedented rate. The future pressures will demand the development and implementation of appropriate methods to address issues of vulnerability to weather and climate. The role of Agricultural Meteorologists is of importance providing the necessary tools to the farmers to overcome the ill effects of climate change long-term climate in addition to combating the extreme weather conditions keep on track the sustainable agricultural development. To meet this objective the Agrometeorologists must also redraw the inputs of Agronomy in addition to other own tools. To achieve this goal the knowledge of the sciences of Agrometeorology and Agronomy are properly understood through most appropriate books. However, a few such books are available as on today.

I am happy that Drs.V. Radha Krishna Murthy, M. Yakadri and P.V.V. Prasad have made sincere efforts to compile and edit the terms in Agricultural Meteorology and Agronomy to bridge the above gap. The authors have also attempted the herculean task of producing simple terms out of more complicated mathematical and statistical equations found in the western books.

I congratulate the authors for writing this valuable book. I am sure that this publication will be useful to the students, teachers, researchers, policy makers, professionals and who might be seeking awareness of the important fields of Agricultural Meteorology and Agronomy. I also feel that the terms given in this book will go a long way in improving the scientific knowledge related to natural resource management, which is the need of the hour across the globe. In addition, I hope that this book will assist the students, researchers, extension workers, planners etc., in their efforts for sucessful application of climate and weather information, data, and forecast systems for agriculature and to assess the strengths, weaknesses, and limitations of the more general use of weather and climate forecast information for agriculture.

S. RAGHU VARDHAN REDDY

Preface

Andhra Pradesh has one of the most variable rainfalls of any other state in India. Climate is a major factor affecting the profitability of rural enterprises in this state. Better knowledge of weather and climate could assist the farmers in the state in making more effective use of resources such as labour, capital, water, etc. However, like thousands of farmers across the globe the farmers in this state are also finding it hard to come to terms with the deteriorating conditions in agricultural sector marked by rising cost of cultivation, and depleting ground water resources. So, for scores of debt ridden small and marginal farmers, there appears to be no respite from the techniques that are already available. The sufferings are unending. They are caught in the vicious circle of mounting debts and failed crops due to weather related disasters. These disasters include drought, untimely rains, hot and desicating winds, hail storms etc. Therefore, the income to the farmers is reducing substantially due to these weather abnormalities/ aberrations. Therefore, there is an urgent need to take up farmer friendly initiatives through better access to weather knowledge that will serve as a non-monetary input. In addition, the knowledge of the farmers should be strengthened to cope with environmental disasters that irreversibly affect their resource base and farming systems. The books which contain terminology that asist the students, researchers, extension workers, planners etc., are most useful and is the need of the hour.

This book is improved over the earlier book "Terminology in Agricultural Meteorology" with additions of new terms from the subjects of Agronomy and related areas. We are confident that the attempt made by us will be an asset to the students of agricultural universities, researchers, extension workers, farmers and others interested in weather and climate related knowledge to solve the field level problems.

We thank Mr. Nikhil Shah, Proprietor of BS Publications for his cooperation in publishing this book.

We are especially grateful to our families for wilfully sacrificing their association with us at homes while we were busy in preparing the contents of this book.

-Authors

Alphabetic Contents

A

Abi

Synonym of *Kharif* season in Telangana region of AndhraPradesh. See '*Kharif*'.

Abiotic environment

This is also known as physical or non-living environment. According to micrometeorologists, this consists of climate, soil, water and air. In this environment a crop derive and respond more vividly for changes in the availability of these components.

Abrasion

The result of erosion by heavy winds on loose soils. The soil looks falter and horizontal without much useful particles.

Absolute acceleration

The rate of change of velocity is acceleration. The meaning of 'absolute' is 'independent'. So, absolute acceleration is the acceleration measured relative to an unaccelerated reference frame.

Absolute cavity radiometer

An instrument used to measure the intensity of direct solar radiation.

Absolute extremes

The highest and the lowest values of a meteorological element especially temperature, that have ever been recorded at a meteorological station. Based on these values normals are calculated.

Absolute humidity

In a system of moist air, this is the actual quantity of water vapour by weight present in the total volume of the system. In other words, this is the ratio of the mass of water vapour in the moist air to the volume occupied by the mixture and is expressed as grams per metre per metre per metre.

Absolute instability

A state of a layer within the atmosphere in which the vertical distribution of temperature is such that an air parcel , if given an upward or downward push, will move away from its initial level without further outside force being applied. The energy transformations take place under this situation.

Absolute scale

The lowest temperature, theoretically possible is known as absolute zero. The zero of thermodynamic temperature is -273.15 degrees Celsius. The absolute scale is the temperature scale based on absolute zero.

Absolute temperature

The temperature recorded from an absolute scale and absolute zero as the base.

Absolute vorticity

The vorticity measured in comparison with a non-rotating frame used as a reference.

Absolute zero

The lowest temperature theoretically possible. This is an hypothetical temperature at which the motion of transmissible particles is totally absent.

Absorption

The process in which insolation is retained by green parts of the plant in shorter wavelengths and by other parts in longer wavelengths. This term has full meaning when the energy retained is irreversibly converted to some other form of energy. Usually natural bodies are rarely, if ever, perfect absorbers.

Absorptive power

Synonym of 'Absorptivity'. The absorptive power of a body is the ratio of radiant energy absorbed to the total amount incident on the same object.

Absorptivity

For an object this is the ratio of the electromagnetic radiant power absorbed to the total amount incident upon the same object. For any object other than a black body, the value is less than one.

Abyssal depth

The depth at which the water remains uniform in temperature, or is stagnant.

Abyssal plain

The flat, gently sloping or nearly level region of the sea floor.

Acceleration due to gravity

The gain acquired in velocity, by a body in space, at a given point, during its free fall. This is equal to 32.08 feet per second at the equator and 32.258 feet per second at the poles.

Acceleration

The rate of change of velocity. Here, it is more appropriate to mention ' time rate' instead of 'rate' because the change is measured always per unit time. However, in all micrometeorological processes any one of these two holds good.

Accessory pigments

These are the pigments which occur in association with chlorophyll. Algae contains these pigments in their antennae. The red algae have phycoerythrins and the phycocyanins are present in blue green algae, where as the carotenoid fucoseanthol is present in brown algae. All these pigments enable the algae to harvest the light quanta over a wider range of wavelengths. When energy in the light quanta is accepted by these pigments then all the interrelated plant physiological processes begin, e.g. photosynthesis.

Accommodation

When an object is to be viewed from a distance the eye should adjust with reference to its focusing power. In this case, the capacity of the eye to adjust itself is known as accommodation.

Accretion

The growth of a body of water or a cloud or a precipitation particle or ice, either due to collision and union of frozen particles or by collection of smaller particles. Both the processes occur only when in contact with supercooled water drops. Among general meteorological terms this is used to indicate the growth of a precipitation particle. The growth of a precipitation particle starts with the collision of a frozen ice crystal or snow flake with a super cooled liquid droplet which freezes upon contact. Like this, a cycle of events take place and precipitation starts.

Accumulated temperature

This is the sum of the departures of temperature from a base or reference temperature. This is useful in the forecast of the phenology and phenophases of a crop under reference. Also referred as 'Growing degree day' or 'Degree day'.

Accuracy

Every instrument has its own degree of precision while measuring a quantity. It is the closeness of the instrument output to the true value of measured quantity.

$$\% \text{ True value} = \frac{\text{Measured value - True value}}{\text{True value}} \times 100$$

Acid rain

The rain formed by reaction of sulphur dioxide and nitrogen oxides from industrial emissions with the moisture available in the atmosphere. These rains are rare, but when occur they cause harm to crop fields. However, these rains may be beneficial when occur on large patches of saline soils.

Actinometer

An instrument used to measure the intensity of radiant energy.

Actinometry

A branch of physics concerned with the study of solar radiation and its different forms.

The measurement of chemical reactions caused by radiation. These are useful in biophysical studies.

Action spectrum

This is the efficiency of photosynthesis by monochromatic light as a function of the wavelength of light. Through this, the photosynthates are formed, of which some are used in respiration and the remaining chemical compounds are stored in different organs of the crop plants.

Active front

During the process of transition between two air masses, if appreciable cloudiness with good amount of rainfall occurs, then the front is considered as active front. These fronts are economically beneficial when occur during the moisture sensitive periods of the crops. Also see 'Front'.

Actual time of observation

As per IMD guidelines, the exact time at which the data on weather elements are recorded. When a barometer is read, it is the actual time of observation in case of surface synoptic observations. As far as upper air observations are concerned the time at which the balloon, parachute or rocket is released.

Adhesion

The effect of attraction produced by dissimilar molecules. This attraction is 'force' and results in sticking to a surface. Example : moisture retension by soil particles.

Adiabatic change

When a change in the pressure and volume of a parcel of air occurs and the heat energy neither enters nor leaves the parcel, then the change is said to be adiabatic.

Adiabatic lapse rate

The rate of vertical decrease in temperature is called the lapse rate. The rate of decrease of temperature with height, of a dry air parcel, when lifted adiabatically is known as adiabatic lapse rate. This is close to one degree Celsius per one hundred metres. Also see 'Adiabatic'.

Adiabatic process

A reversible thermodynamic change of state of a system in which there is no addition or removal of heat or mass across the boundaries of the same system. This is same as 'Adiabatic'.

Adiabatic

A thermal process is called as an adiabatic process when the heat energy neither enters nor leaves the system in which the process is happening.

Advection fog

The main types of fog are 'air mass fog' and 'frontal fog'. The advection fog is a sub-type in 'air mass fog' and is formed because of advection of warm air over a colder surface. Also see 'Advection'.

Advection

The horizontal transfer of heat energy, effected by horizontal exchange of air, because of the large scale motions in the atmosphere.

Advisory

Statements that are issued by the **National Weather Service** for probable weather situations of inconvenience that do not carry the danger of **warning** criteria, but, if not observed, could lead to hazardous situations. Some examples include farming advisories stating possible attacks by pests and diseases.

Aeration of soil

The God created organisms not only on the surface of the Earth, but also inside it upto some layers. So, they need oxygen for survival, the concentration of which will never be zero, by volume inside the soil.

Same is the case with carbon dioxide, which will never go above five percent and below one percent. This is a clear indication that there is aeration in the soil and to the crop plants, this aeration is improved regularly by tillage operations.

Aeration

The process by which air and other gases in a medium are renewed or exchanged. For instance, summer ploughings in crop fields are done to provide the soil with the best possible air to reap better harvest.

Aerial perspective

A visual clue to relative distances, in which objects loose detail and their colour shift towards hazy blue according to the distance between them and the eye. In remote sensing of large crop fields this term has greater significance.

Aerial photograph

A vertical or oblique photograph taken from the air. Photographs taken in this way will give a bird's eye view of the whole crop field in which an experiment is going on.

Aerial seeding

In hilly areas and on forestlands, if new species of plants are proposed to be grown, the seeds are dropped from aeroplanes, helicopters, etc. It is to be noted that, this is usually done in areas where seedbed preparation is not possible. Usually, this process is taken up when all the climatic parameters are congenial, in addition to the moisture availability in that region.

Aerodynamic roughness

Most of the drag on rough natural surfaces such as forests and crops is due to form drag. This arises from the deceleration of fluid in front of an element normal to the flow. If the shear stress at the surface is mostly because of this form drag of the rough surface and the viscous shear stresses are negligible in comparison, then it is said to be aerodynamic roughness.

Aerodynamically rough surface

A surface is said to be aerodynamically rough when the roughness elements are sufficiently large such that the turbulent boundary layer reaches the surface. The form drag and viscal forces also play a role in the process.

Aerodynamically smooth surface

An opposite of aero-dynamically rough surface, which means the roughness elements are significantly small to be entirely embedded in the laminar sub-layer.

Aerodynamics

As far as micrometeorological studies are concerned, this is the study of the effect of air and other gases in motion, on the soil and crop surface. In general and in broader sense, this is the study of the effect of air on the motion and control of solid bodies like aircrafts, satellites, etc.

Aerology

The study of the Earth's free atmosphere and space beyond.

Aerosol

A collection of microscopic liquid or solid particles dispersed in the atmosphere. These are larger in number in the low troposphere.

AFOS

Acronym for Automation of Field Operations and Services. It is the computer system that links National Weather Service offices together for weather data transmission.

After image

A visual effect occurring after a light stimulus has been removed. This effect can be either positive (when the brightness or colour is the same as the stimulus) or negative (when the brightness or the colour of the image is the opposite of the stimulus). In the crop fields, this term is used to indicate the state of the crop after sunset. In diurnal pattern studies this is a common usage.

Agricultural climatology

A branch of applied climatology to study the influence of climate on different processes of agriculture. The abbreviated term is Agroclimatology. Also see ' Agroclimatology'.

Agricultural drought

This is a situation which is a result of inadequate rainfall. Because of this, the soil moisture falls short to meet the demands of the crop during its growth. Since, the soil moisture available to a crop is insufficient it affects growth and finally results in the reduction of yield.

Agricultural meteorology

- A branch of applied meteorology which is concerned with the study of meteorology in relation to agriculture.
- A science investigating the meteorological, climatological and hydrological conditions which are significant for agriculture owing to their interaction with the objects and processes of agricultural production.
- A study of those aspects of meteorology which have direct relevance to agriculture.

Agriculture

The entire complex of Trees/Crops/Animals and other microflora and fauna in a geographic location together with overall environment that is modified/changeable by agricultural practices.

Agrisilviculture

A self-sustaining land use system in which agricultural crops are combined deliberately with trees.

Agrobiology

A branch of agronomy that deals the relationship between yield to the added or available nutrient under study.

Agroclimate

The combined influence of climatic elements that make possible the cultivation of crops. The point to be stressed in this regard is the totality of the behaviour of these elements including their intensity and frequency not only on plants and animals but also on the soil medium.

Agroclimatic index

As the agricultural sciences are developing with the passage of time and also due to the advances taking place in the technology, so many indices are being developed to express the relationship between climate and agricultural production. Of them, the agroclimatic index is the one which expresses this relationship in quantitative terms.

Agroclimatic region

A region in which all the climatic parameters are almost homogenous as far as agricultural crops and their production is concerned.

Agroclimatology

Abrreviated form of Agricultural climatology. This is a branch of science which deals with the relationship of climatic regimes and agricultural production including instrumentation and weather forecasting. Also see 'Agricultural climatology'.

Agroecological zone

Climatically and edaphically a large extent of area in which a particular crop expresses the more or less same yield potential.

Agroecology

A branch of ecology that deals with reciprocal relationship between agricultural crops and their environment.

Agroenergy

A term used in agriculture about requirement and yield of energy. Single most source of energy for crop plants on earth planet is sun. This energy in the form radiation is captured and converted into usable energy like food, fodder, feed and fuel through an Anabolic process called photosynthesis in turn these could be transformed into other forms of energy like gaseous or liquid forms.

Agroforestry

A self-sustaining land management system which combines production of agricultural crops with that of tree crops and also with that of the live stock simultaneously or sequentially, on the same unit of land. The by-product of one enterprize meets requirement of other forming a mutually inclusive forms.

Agrology

Applied phases of soil science and soil manage in agriculture.

Agrometeorologist

An agricultural scientist who applies relevant meteorological skill to help the farmer and agricultural workers to make the most efficient use of physical environment for improving agricultural production both in quality and quantity. He has to put the science of meteorology to the service of agriculture.

Agrometeorology

An abbreviated form of agricultural meteorology. See 'Agricultural meteorology'.

Agronomy

There are many definitions. However, the central theme of all attempts to define revolve around and majority accept that, it is a subject of agricultural sciences in which the biological, chemical and physical knowledge is integrated and applied into useful crop production systems.

- The science of the laws of field cultivation with a view to potential or actual production.
- A branch of agriculture dealing with field crop production and soil management.

Agronomic Management Index (AMI) :

A relative yield index on part of weed control that is contributing to Agronomic efficiency and is given by the formula,

$$AMI = \frac{\text{percent yield - percent control of weeds}}{\text{percent control of the weeds}}$$

Agrosilvipastoral systems

A branch of agroforestry that sustains land for concurrent production of agricultural crops, pastures and tree components for domesticated animals.

Agrostology

A science of classifying grasses, and their indepth study – of management and utilization.

Air currents

The columns of air moving in vertical direction, in the atmosphere.

Air density

The mass density of air in terms of weight per unit volume. The density of air plays an important role with reference to the water vapour content and its dynamics in the atmosphere as well as in crop canopies.

Air drainage

If the density of cold air increases because of the presence of radiating surface and effect of lapse conditions, the same air is transferred downhill. The gravity forces acting on the air make this possible, as the air is already under the influence of advection.

Air mass classification

A system used to identify and to characterize the different air masses according to a basic scheme. The systems most commonly used classify air masses primarily according to the thermal properties of their source region like tropical (T), 'polar (P)', etc. They are further classified according to moisture characteristics as 'Continental (C)', or maritime (m). Further, there are several classifications in use which are widely accepted.

Air mass source region

The air masses are classified according to some basic schemes. The climatology of air masses deals with the representation of climate of a region by the air mass frequency and the characteristics under which it lies. When the bodies of air remain sufficiently long over an extensive area on the surface of the earth they acquire the temperature and moisture properties of that region. The areas over which they exist are called their source regions.

Air mass thunderstorm

A thunderstorm that is produced by convection within an unstable air mass through an instability mechanism.

Such thunderstorms normally occur within a tropical or warm, moist air mass during the summer afternoon as a result of afternoon heating. They dissipate soon after sunset. Such thunderstorms are not generally associated with fronts and are less likely to become severe than other types of thunderstorms. However, brief heavy downpours may occur due to these air mass thunderstorms.

Air mass

A synoptic scale body of air covering an areas of about eighteen hundred kilometers or even more with thickness upto several kilometers. The physical properties are related to geographical source, region, and the temperature and humidity are almost uniform.

Air parcel

A unit body of air. Here 'unit' is a relative term, in sense, a small body or a medium body, etc., as compared to its actual existence in a given situation. But, this small body should be in function during a physical process. To be in function the body must be uniform, atleast, to be identified during a process like buoyant thermal or adiabatic change in air, etc.

Air pollution

Air is considered as polluted when it is changed in its quality and composition as a result of the activities of man. This becomes an ecological problem when it reaches above tolerant levels for crop plants.

Air quality standards

The maximum level which will be permitted for a given pollutant. Primary standards are to be sufficiently stringent to protect the public health. Secondary standards must protect the public welfare, including property and aesthetics.

Air temperature

The temperature recorded on a thermometer after exposing the same to air by protecting it from direct solar radiation with a shelter.

Air

A colourless, tasteless and odourless mixture of gases which surround the earth and extend to a height of several hundred kilometers is known as atmosphere. The gaseous component of the atmosphere, particularly that exists in the troposphere is called as 'air'. Air is a mechanical mixture of gases and its density decreases with altitude.

Airmeter

An instrument used to measure the speed of the wind, by taking into consideration not only the exposure but also the number of linear metres the air passed through the instrument.

Albedo

This is an astronomical term for 'whiteness' and in practice the value is applied primarily to solar radiation. When the incident solar radiation beam passes through the atmosphere, it is reflected back into space by different bodies like clouds, ice, water, dust, trees, plants and the land itself. The ratio (percentage) of the amount of reflected back electro-magnetic radiation by a body to the amount of incoming electro-magnetic radiation incident upon it is referred to as albedo. Here, the 'amount' means the complete wavelength range of solar radiation. However, depending upon the use and interpretation, sometimes only the visible part (0.4 to 0.7 microns) of the spectrum is taken into account. For meadows and fields the albedo (percent) is 12-30, for fresh snow cover 75-95 and for dark cultivated soil 7 to 10. Albedo is influenced not only by the nature of the surface but also by its moisture content at any time. Wet surfaces appear darker than dry surfaces and in general, the albedo value decreases when a body becomes wet.

Alimeter

An instrument used to measure the altitude above a given level.

Alkali soil

problematic soil usually with pH above 7.5.

Allelopathy

The release of some chemicals through roots or leaf, stem exudates or decaying plant residue having harmful effect on, other organisms.

Alley cropping

A farming system in which arable crops are grown in alleys formed by trees or shrubs, established mainly to hasten soil fertility and soil productivity.

A sustainable agroforestry system where in the short duration grain crops like pearl millet, sorghum, cowpea etc., are grown in between the well managed, soil fertility enhancing perennial hedgerows spaced at regular intervals. The hedgerows formed with trees like, subabul *Glyricidia sp* are kept pruned during cropping to reduce competition with crops. It is otherwise called as "hedgerow inter cropping".

Alluvial soil

A soil developed due to deposition of clay and humus called alluvium that did not exhibit any horizon or modification.

Alpha radiation

A kind of ionizing radiation in which alpha particles are given off. These rays have less wavelengths and higher energy.

Alpha ray

A stream of alpha particles. The ray theory relies more in explaining these particles as a stream of photons.

The property of spontaneous disintegration possessed by certain unstable types of atomic nuclei is known as radio-activity. The disintegration is accompanied by emission of either alpha, beta or gamma rays. A Helium nucleus with a charge of plus two and a mass of four atomic units which is emitted by its nuclei is an alpha particle.

Alternating current

A flow of electricity in which the electrons reverse their direction periodically.

Altimeter setting

The **pressure** value to which an aircraft **altimeter** scale is set so that it will indicate the **altitude** above **mean sea level** of an aircraft on the ground at the location for which the value was determined.

Altitude

The angular distance (height) of a place or an extra-terrestrial body from the mean sea level. Sometimes the distance of a body is measured from the horizon along the meridian passing through it.

Altocumulus

These are middle level clouds and the mean height is 2 to 7 kilometers. The prefix 'altum' refers to height and 'cumulus' means 'heap'. These are white or grey heaps of clouds, appear fibrous and diffuse, and sometimes referred to as 'sharp clouds' because of high globular structure.

Altostratus

These are middle level clouds and the mean height is 2 to 7 kilometres. The prefix 'altum' refers to height and 'stratus' means layer or sheet. These are greyish layers with fibrous appearance covering the sky almost completely and halos are never seen. Precipitation falls occasionally either as drizzle or snow.

Aman rice

A direct seeded rice on dry soil in premonsoon period in deep water areas (1-6 meter) like in Assam, West Bengal, Bangladesh. Rice plants grow along with rising water level and subsides as water recedes. Cultural operations are different compared to a traditional low land rice.

Ambient temperature

The meaning of ambient is the preceding or surrounding a phenomenon. In the same way, ambient temperature refers to the temperature of the surrounding atmosphere, water and soil. Usually, the dry bulb temperature is considered as ambient temperature in crop-weather studies.

Amendment

A chemical substance that is added to improve the soil. Materials such as lime, gypsum, sawdust, sulphur or soil conditioner that are added to improve the physico-chemical properties of the soil.

Ammonifying bacteria

Protein and other nitrogenous substances are decomposed by a group of soil bacteria to release ammonia.

Ammeter

An instrument used to measure the intensity of electricity in a circuit.

Ammonia volatilization (soil)

Loss of nitrogen in the form of gaseous ammonia (NH_3) from soil when ammonical fertilizers applied to soil containing appreciable amount of $CaCO_3$ under high pH and temperature conditions. Ammonia also liberated from plant canopies some times depending upon the influence of factors. Ammonia readily identifiable product in nitrogen mineralisation, is formed continuously in the soil and water of flooded paddy. Organic matter decomposition in absence of oxygen and high pH favours ammonia volatilization from flooded soil.

Ammoniated superphosphate

Obtained when superphosphate is treated with ammonia or with solution which contains ammonia or other compounds of nitrogen.

Ammonium fixation

Adsorption of ammonium ions by soils or clay minerals in such form that they are neither water-soluble nor readily exchangeable.

Ammonium phosphate

When phosphoric acid is treated with ammonia a product is obtained that consists monoammonium phosphate and di-ammonium phosphate or both.

Ammonium sulphate nitrate

A double salt of ammonium sulphate and ammonium nitrate containing 26 percent of nitrogen.

Ammonium sulphate

An acid forming fertilizer obtained by treating ammonium salt with sulphuric acid containing 20% of nitrogen.

Ampere

The unit of current intensity. Also, the rate of transfer of electric charge equivalent to one coulomb per second.

Amplifier

A device for increasing the strength of a signal or energy of a circuit.

Amplitude

In an oscillatory motion of a quantity this is the amount of displacement. This is applied to the motion of an air particle displaced by a sound wave. In case of a pendulum the amplitude is half the length of the swing.

An Einstein.

The avogadro's number of quanta, equivalent to 6.023×10^{23}. Named after Albert Einstein (1879-1955).

Anabatic wind

An up-slope flow of air or wind due to local surface heating. This is a synonym of valley breeze and sometimes accompanied by the formation of cumulus clouds.

Anaerobic respiration

Incomplete oxidation of foods in living cells with release of energy in absence of atmospheric oxygen.

Anafront

A front which has ascending air at one side, particularly one experiencing the unusual phenomenon of rising cold air. Such fronts have local influences under natural conditions.

Analytical model

The economic yields of crops are influenced by weather and soil management, in addition to the characteristics of the crop itself. In analytical models, all relationships are expressed in closed form, so that the equation can be solved by the classical methods of analytical mathematics.

Anemograph

An instrument used to measure both velocity and direction of the wind.

Anemometer level

The level at which an anemometer is exposed to record the observed values.

Anemometer

An instrument used to measure the wind velocity.

Aneroid barometer
An instrument used to indicate the atmospheric pressure.

Angle of incidence
The angle that a light ray falling on a surface makes with the a line normal to the same surface. Here 'normal' means 'perpendicular'.

Angle of intersection of two curves
The angle of intersection of two curves at a point of intersection is the angle between the tangents to the curves at that point.

Angle of reflection
The angle between a ray of light reflected from a surface and the normal to the surface at that point, which means, this is the angle a reflected wave makes with the reflecting surface.

Angle of refraction
When a ray of light travels obliquely from one medium to another it is bent at the surface separating the two media which is known as refraction. The refraction occurs because light travels at slightly different velocities in different media. The angle that the direction of travel of refracted wave makes with normal to the boundary is known as angle of refraction.

Angstrom
A unit of length usually associated with the measurement of light or other waves and intra-molecular distances.

Angular acceleration
The change of angular velocity with reference to a unit time.

Angular momentum
The vector product of the momentum of a body and its radial distance from the axis of rotation. Also known as moment of momentum.

Angular velocity

The time rate of change of the direction of a body about an axis. This is expressed in angle turned per unit time.

Anhydrous ammonia

An ammonical source of nitrogenous fertiliser containing 82% nitrogen. Highly acidic in nature, can be stored under standard conditions need special equipment to apply in the field.

Annidation

One type of interaction where one species is complementary in supplying the growth factor to the other species. Example in a legume + cereal association a part of nitrogen fixed by legume will be supplied to non legume may be directly or indirectly.

Annual range of temperature

The difference between the warmest and the clodest monthly means in a given areas or locality.

Antarctic Ocean

The Atlantic, Pacific and Indian Oceans that reach the Antarctic continent on their southern extremes. However, this is not officially recognised.

Antarctic

On the earth the area south of $66° 32'$.

Antecedent moisture

The soil moisture condition in an area before a rainfall, which moderates the run-off response of an area to the rainfall. This moisture has large bearing in the profile moisture calculations of different cropped fields.

The degree of wetness of the soil at the beginning of a run-off period.

Antecedent precipitation index

An index of moisture conditions in a catchment area used in the preparation of river level forecasts. This term is used to assess the pre-storm basin moisture condition on the basis of rainfall data. In other words, this is a weighed sum of preceding rainfall within time units. This index is also useful in scheduling of irrigation requirements of different crops.

Anthesis

Opening of flower buds in an inflorescence for pollination as well as fertilisation. The duration may vary from 6 to 12 hrs preferably morning to afternoon that may occur either top to bottom or vice versa.

Anticyclone

A synoptic scale weather system with winds blowing clockwise in the northern hemisphere and counter clockwise in southern hemisphere. An anticyclone is a warm sinking air. The wind flows out from areas of high pressure in which air does not rise enough vertically. Rainfall is not common in areas of high pressure.

Antitrade

A westerly air current. In both the hemispheres, particularly in sub-tropical regions these winds blow above the trade winds.

Antitrades

A deep layer of westerly winds in the troposphere above the 'surface trade winds', which represents the upper limits of the Hadley cell, within which occurs the poleward transfer of heat and momentum and water vapour. The meteorologist consider these antitrade as energy distributors, when they exist on large areas.

Anvil

The ice portion of the cumulonimbus. Usually this is seen at the top of this cloud.

Aperture

The size of the opening admitting light in radiation measuring instruments or in any optical instruments.

Aphelion

The time at which the Earth and Sun are at a maximum distance between them.

Apogee

The point farthest from the earth on the moon's orbit. This term can be applied to **satellites** also.

Apparent density of soil

The ratio between mass of all components to the total volume of all components of a soil. This is expressed in grams per centimetre per centimetre per centimeter.

Apparent gravity

The force of attraction between the Earth and a body on its surface or within its gravitational field is known as gravity. The gravitational acceleration measured relative to a fixed point on the surface of the Earth is known as apparent gravity.

Apparent specific gravity

It is nothing but bulk density i.e., oven dry weight of the soil per unit volume expressed as gr/cc. The value varies in between 1.4 to 1.6.

Apparent wind velocity

The vector difference between the true winds and velocity of the object.

Applied climatology

The branch of climatology to study the applications of scientifically analysed climatic data for any operational purposes, like developing technology for dry land agriculture, etc.

Aquatic

Also called as hydrophytes that need submergence or partial submergence of water for their normal growth.

Arable crops

Crops which need cultivation packages like tillage, sowing, fertilization, weeding, irrigation, harvesting etc.

Arable farming

Farming systems that require tillage of hoeing for production of crops.

Arable land

Cultivable land that is utilised for production of crops that are free from natural vegetation like grass etc.

Arboriculture

Growing of woody plants with a purpose of decoration and shade.

Archimedes principle

The buoyancy equals displacement. For example, when a body is submerged in a fluid the buoyant force on the submerged body is equal to the weight of the displaced fluid.

Arctic air mass

An air mass that developed around the Arctic. It is characterized by being cold from surface to great heights. The boundary of this air mass is often defined by the Arctic front which is a semi-permanent, semi-continuous feature.

Arctic front

The surface of discontinuity between very cold (Arctic) air flowing directly from the Arctic region and another less cold and, consequently, less dense air mass. In the Arctic region of the world, these have a large bearing on the climate.

Arctic

The area to the north of latitude at sixty six degrees and thirty two minutes, on the Earth.

Area-time equivalency ratio

Ratio between number of hectare-days required in sole cropping to number of hectare days required in inter cropping to produce the similar yields of either of component crops.

Argon

The colorless, odourless inert gas that is the third most abundant constituent of dry air, comprising 0.93 % of the total by volume and 1.286% by weight.

Arid climate

An extremely dry climate in which desert type of vegetation is found, where the ratio of precipitation to evaporation is the lowest. Summers are comparatively dry with small rainy season

Arid zone

A desert climate zone, in which desert vegetation is noticed and precipitation is much lesser than evapotranspiration demands of plants.

The portion of the Earth between latitude $15°$ and $30°$ north of the equator and the same latitudes in south, which consists of most of the world's warm deserts. In practical usage, this term is applied to areas on the Earth which receive less than 250 millimetres of rainfall annually. Most of the arid zone crops have shorter growth phases as compared to temperate crops.

Aridity index / Aridity factor

An index (number) proposed by Thornthwaite in 1948 and subsequently by several authors in different contexts. This is an indication to determine the degree of dryness of a climate as a function of various climatic factors, such as temperature, heat, moisture, potential evapotranspiration, etc.

This is a characteristic of a climate relating to insufficiency of precipitation to maintain vegetation. So, this is a measure of effectiveness of the precipitation and evaporation. More the aridity index more is the evaporation that occurs as compared to the rainfall received. For example, if the aridity index of Hyderabad in India is two (2), the Sun would evaporate twice the amount of rainfall received at this place.

Artificial fertilizers

Commercially manufactured or mineral product that supplies one or more essential plant nutrients that are readily available also known as inorganic or mineral fertilizers.

Arviculture

A synonym to crop science.

Ash

A product that results after complete burning of organic matter, non volatile in nature and rich in potash.

ASOS

This is the acronym for Automated Surface Observing System. This system is a collection of automated weather instruments that collect data. It records, surface based observations from places that do not have an observer,

Astronomical twilight

The time after nautical twilight has commenced and when the sky is dark enough, away from the sun's location. This helps in astronomical work to proceed.

Atmometer

An instrument used to measure the evaporation rate.

Atmosphere

The mass of air surrounding the Earth. Nearly half of the atmosphere lies upto 5 km from the surface.

Atmospheric boundary layer

The low tropospheric portion of the atmosphere, which is also known as planetary boundary layer.

Atmospheric circulation

The complete or partial gaseous motion in the envelop of the Earth's surface.

Atmospheric counter radiation

See 'Down-ward terrestrial radiation'.

Atmospheric disturbance

The disturbance caused to the meteorological conditions at any place, due to a minor depression, because of which there are chances for the development of cyclonic circulation.

Atmospheric pressure

The term pressure means, the force per unit area acting on a surface. The pressure exerted by the atmosphere is known as atmospheric pressure and is usually measured in millibars. On an average, the atmospheric pressure at sea level is recorded as 1013.2 millibars, or 10^6 dynes per centimetre per centimetre.

Atmospheric radiation

A part of the terrestrial radiation emitted by different components and constituents of the atmosphere.

Atmospheric waves

Air is a fluid. It moves in short rain and sometimes solitary in the form of waves. In the movement, undulations of horizontal or vertical nature are formed. These are called as atmospheric waves.

Atmospheric window

In the spectral region between 8.5 and 11 microns only a small portion of the long wave terrestrial radiation is absorbed by the atmosphere. Because of its inability to retain the energy in this segment of the wave band it provides 'windows' through which the energy ultimately escapes into the space, which are known as atmospheric windows.

Atmospherics

Disturbing effects produced in radio receiving apparatus by atmospheric electrical phenomenon such as an electrical storm. Because of these atmospherics low and high frequency sound effects are noticed.

Atom

In an element, this is the smallest part that takes part in a chemical reaction.

Attenuation of solar radiation

In crop canopies and forest plantations the solar radiation beam suffers loss of energy by differential depletion while passing through them. These losses are due to selective absorption by different plant parts particularly leaves and to some extent scattering by some constituents of the plant atmosphere. This loss of energy is called as attenuation of solar radiation.

Attenuation
When a radiation beam passes through an absorbing and scattering medium, the intensity of the radiant energy relatively decreases. Here, 'relatively', means the difference of intensity between, before passing through the medium and after passing through the medium. What exactly happens is the 'loss of power' in the radiation, and 'medium' means, a crop canopy, as far as micrometeorological measurements are concerned.

Aurora
It is created by the radiant energy emission from the sun and its interaction with the earth's upper atmosphere over the middle and high latitudes. It is seen as a bright display of constantly changing light near the magnetic poles of each hemisphere. In the Northern Hemisphere, it is known as the aurora borealis and in the Southern Hemisphere, as aurora australis.

Autecology
Interaction of an individual organism with its environment.

Autumn
This refers to the months of September, October and November in the North Hemisphere and the months of March, April and May in the Southern Hemisphere. This is the period between the autumnal equinox and the winter solstice, which is characterised by decreasing temperatures.

Available effective rainy period
The week for taking up sowings after receiving sufficient rains, the word 'sufficient rains' has great significance as the same has to be derived based on statistical and mathematical procedures, by using standard formulae and equations.

Available nitrogen
Theoritical field available nitrogen to NO_3 from potentially available source in the rhizosphere or nitrogen some times may converted to NH_4^+ that adhered to micelle or may be utilized by some plants like rice in early stages.

Available potential energy

The energy which a body possesses by virtue of its position is called as potential energy. When the potential energy is converted into kinetic energy it is called as available potential energy. The condition to be fulfilled is that the same should happen by realistic redistribution of air parcels in the atmosphere or a crop canopy.

Available water holding capacity

The amount of water available in the soil between field capacity and the permanent wilting point. This varies with the structure, texture and composition of the soils.

Available water

That part of the soil water in between fields capacity and permanent wilting point and is absorbed by plant roots against a pressure of -0.3 to -15 bars for normal metabolic activities.

Available water

The part of the water in the soil that plants can absorb. This water is effectively taken by plants for use in their physiological processes.

Avalanche wind

A wind, often destructive over a distance, produced by an avalanche. These winds, when accompanied by a large scale snow cause chilled weather over large areas.

Average

In Meteorology and Agrometeorology, this term is often used to refer to the experience of climatic elements over an area for a period of thirty years. When such a longer period is taken, the weather elements like rainfall, temperature, humidity, wind velocity, direction , etc., shall have to be taken into account as average values. Even in the numerical weather prediction average weather elements recorded for a period of thirty years is taken into consideration for calculations.

AVHRR

Acronym for Advanced Very High Resolution Radiometer. It is one of the main sensors in the satellites for remote sensing purposes, being used across the globe.

Avogadro's number

The Avogadro's law states that, equal volumes of all gases contain equal number of molecules at the same temperature and pressure. The number of molecules are close to 6.023×10^{23} in a mole and this is the Avogadro's number.

Avogardro's hypothesis

For two different gases, if volumes are equal, the number of molecules present in them are equal, at the same temperature and pressure.

AWIPS

Acronym for Advanced Weather Interactive Processing System. It is the computerized system that processes the weather data for forecast offices. This system helps in scientific way of planning the forest development.

Axillary tiller or bud

The tiller or bud produced from the junction of leaf and stem.

Azimuth

The angular distance from the north and south point of the horizon to the foot of the vertical circle through a heavenly body. The azimuth of a horizontal direction is its deviation from the north or south. The wind direction specified by the angle in degrees from which the wind is blowing, counting clockwise from zero to true north is also called as the azimuth.

Azimuthal angle

This is the angle between true north and the projection of the Sun's rays on to the horizontal. Often, this angle is taken into account to calculate the flux of radiation over an area.

Azolla

A group of aquatic ferns capable of fixing high level of nitrogen from the atmosphere and are widely grown as a bio-fertilizer crop in low land rice cultivation systems. Anabaena, a blue green algae that fixes atmospheric nitrogen in symbiosis with fern. Two to three life cycles of fern are completed in single season of rice that enriches the soil fertility. Some azolla are used as feed for milch cattle.

Azores high

A semi-permanent, subtropical area of high pressure in the temperature gradient along the East Coast of the United States during the winter that gives rise to intense cyclogenesia.

Azotobacter

A non-symbiotic bacteria that fixes atmospheric nitrogen with help of carbohydrates that is aerobic but susceptible to phosphate deficiency.

B

Back furrow

It is customery to convenient to start ploughing from centre to side, so that unploughed land is not left in the corners. A ridge left at centre of the land while ploughing is started is called back furrow.

Background radiation

The radiation produced by cosmic rays together with the natural radioactive isotopes from soil and water. This is viewed as a major component in the mineral absorption studies of crop plants.

Backing

A change of wind direction in an anti-clockwise direction. In other words, when the wind direction is changing azimuth is decreasing.
The wind that shifts its direction in an anti-clockwise direction.

Backscatter

The energy reflected or scattered by a target. This term is often used in radar systems. In the recent past, the radar technology is being used extensively in the medium range weather forecasting.

Bad weather

Sky with non-clearing thick low clouds.

Bagasse

A by – product from sugar cane industry that is obtained after crushing the stalks for juice extraction. It is used in different ways like correcting the alkaline soil, or for generation of electricity, cardboards etc.

Balanced fertilization

Application of nutrients to soil in a suitable proportion through fertilizers for proper growth of plants. For example, if the NPK status of the fertilizers is in the ratio of 4:2:1 then fertilisation is said to be a balanced one.

Band placement

A method of fertilizer application for less mobile nutrients to one inch below and two inches away from the seed.

Banner cloud

A banner like cloud streaming off from a mountain peak. This is also called as cloud banner.

Bar

A unit of pressure, (usually the atmospheric pressure) and one bar is equivalent to a pressure of one million dynes per centimetre per centimetre; one hundred kilopascals; 29.53 inches or 750-760mm millimetres of mercury at zero degrees Celsius.

Barn yard

Barn means an area enclosed by fencing to stalk the animals and area adjacent to this barn is called barn yard.

Baroclinic instability

The instability state of the atmosphere in large baroclinic surfaces and their associated thermal winds. A consequences is the cyclogenesis.

Baroclinic

The state of the atmosphere in which temperature gradient exists in isobars. In other words, the atmosphere has isotherms which are not parallel to the isobars, which means that there is a temperature gradient along the isobars. The result of this condition of the atmosphere is a variation of the geostrophic wind with height.

Barograph
A self-recording barometer with a graphical output.

Barometer
An instrument for measuring the atmospheric pressure. The main types are aneroid and mercurial barometers.

Barometric pressure
The pressure exerted by the atmosphere at a given point.

Barometric tendency
The change of barometric pressure within a specified period of time. In aviation weather observations, routinely and periodically determined for a 3 hour period. However, in the daily recording of weather data, at the time of observation, this tendency is recorded. Whenever, a storm is active, a periodical and cyclic tendency need to be observed and recorded.

Barotropic
A fluid state in the atmosphere, in which horizontally uniform temperature exists at all heights, the consequence of which is the constancy of the 'geotropic' wind with height.

Barren land
Land that is deprived of vegetation.

Basal dose
That part of the total quantity of fertilizers or manures applied to the soil just before or after the crop is sown.

Base (period)
Number of days during which irrigation is supplied to a crop.

Basic slag
A by-product of steel industry obtained from phosphate iron ores. It contains about 6 to 18 percent phosphoric acid.

Basin irrigation method
A surface irrigation method in which water is applied in small round plots formed by ridges. In this two types. 1. Ring basin 2. Check basin.

Basin listing
Ploughing with ridger that makes furrows at regular intervals for insitu rain water conservation, furrows at regular intervals to create small basin to capture and store rain or applied water.

Bathyhermograph
The instrument used to obtain a record of temperature against depth (pressure) in the ocean. May be referred to as a B.T.

Baul unit
Unit of fertiliser or any other growth factor, be taken as that amount necessary to produce a yield of 50% of the maximum possible. The values of baul unit in pounds per acre of N, P, K are 223, 45, 76 respectively.

Beam
The part which transmits the power of the animal to the plough.

Bearded
Having an awn or bearing long hair, which is extension of lemma and palea.

Bed planting
Sowing on flat-topped ridge which is elevated and separated by furrows on either side for proper drainage, irrigation.

Beer's law
This law states that, the transmission of incoming short-wave radiation into a plant canopy shows an approximately exponential decay with the depth of penetration. In other words, the relative radiation intensity decreases exponentially with increasing leaf area. The theory indicate that the atmospheric envelop of gases surrounding the Earth absorbs considerable portions of the solar beam. The attenuation is a function of the constituents of the atmosphere and because of selective absorption by these constituents, certain wavelengths are more sharply affected than others. This law clearly describe the reduction in flux density of light beam, as a function of depth of homogenous absorbing medium.

Bel

Ten (10) decibels is a bel, and also the logarithmic expression of the ratio of two quantities.

Bellani spherical pyranometer

An instrument used to measure the solar radiation falling on a spherical surface from the hemispherical sky and also that terrestrial radiation which is reflected from the ground.

Bellot winds

Refers to the winds that blow in the Canadian Arctic through the narrow Bellot Strait between Somerset Island and the Boothia Peninnsula.

Benchmark survey

A systematic study aimed at collecting data, e.g. on existing crops, varieties, yields, socio-economic constraints before a project begins. Data collected depict the existing picture of the survey areas with regard to selected parameters and can be used to evaluate the results of the project.

Benefit : Cost ratio (B : C ratio) :

An important economic indicator that tells about profitness of a crop, cropping system or farming system.

It is given by the formula

$$\frac{\text{Net returns}\left(\text{Gross returns - Cost of cultivation}\right)}{\text{Cost of cultivation}}$$

It is also called as net returns per rupee investment. Generally b : c ratio of 2 or more than two is considered as most acceptable one.

Bergeron-Fin deisen process

An initial stage of precipitation by preferred growth of ice crystals over the water droplets in cloud comprising of super cooled water and ice.

Bermuda high

A semi-permanent, subtropical area of high pressure in the North Atlantic ocean that migrates east and west with varying central pressure. Depending on the season, it has different names. When it is displaced westward, during the Northern Hemispheric summer and fall, the center is located in the western North Atlantic, near Bermuda.

Bernoulli's principle

At any point in a tube, through which a liquid is flowing (or is at a constant level) the sum of pressure energy, kinetic energy and potential energy is constant. In other words, the velocity and pressure are inversely proportional to each other and vice-versa. Here, a very important point any micrometeorologist should remember is that the fluid should be incompressible. This principle can be tested in water movement near root zones of any crop.

Beta plane

An electron or positron emitted by a radioactive nucleus is called a beta particle. However, there is no relation between a beta particle and the beta plane. Beta plane is an entry in vorticity equation in establishing the relation with coriolis parameter.

Beta rays

Beta rays are the flow of beta particles. Beta rays are superior to alfa rays as far as penetrating power is concerned and are emitted with more velocities than light.

Beufort wind scale

The scale devised by Admiral Sir Francis Beufort in the year 1805, for estimating and reporting wind speed. In this scale, to measure the speed of wind, a series of numbers are given from 1 to 12. The number on the scale can be calculated easily by the effect of wind strength on surface features such as trees and sea-state.

Binder

Mechanical harvesting where crop is cut and made in to bundles.

Biosoils

Muck or peaty soils developed under swampy marsh or mangroove type of vegetation or under humid or subhumid regions of soils with a muck or peaty surface underlain by peat.

Bio-assay (biological assay)

Determination of the relative strength of a substance (hormone, drug, insecticide or herbicides etc.) by comparing its effect on a test organism with that of a standard preparation.

Bioclimatics

Science that deals with response of plants influenced by climatic factors, including geography, elevation and altitude upon plant responses.

Bioclimatology

This is the study of the relationship between climate and living organisms, like human beings, animals, crops, forests, etc. In broader sense, this field of science brings together many sciences related to climate and ecology, and study their relationships for considerably longer periods.

Bio-conversion

Conversion of one form of energy into another by using living organisms (Animals, plants or micro-organisms). Example Photosynthesis.

Bio-energy

Energy obtained from living sources.

Biogas

An ecofriendly way of producing the energy (gas) from domestic farm wastes by subjecting them to anaerobic fermentation in the presence of methane forming bacteria that contains methane, carbondioxide, hydrogen etc.

Biological control (pests)

Controlling of insect pests and weeds by using the predators, parasites and pathogens.

Biological yield

Total day matter produced by plant that includes grain yield plus straw yield or total forage yield in fodder crops.

Biomass

The total quantity, at a given time, of living organisms of one or more species per unit of space (species biomass) or of all species in a community (community biomass). However, in micrometeorology, this term is always used to indicate the extra growth that the crop acquired from the previous date of observation. In practice, biomass increase is always related to radiation use efficiency.

Biome

A vast community of land which is characterized by its flora and fauna. In crop weather studies, a crop grown in a watershed is also denoted by this term.

Biophysics

The study of phenomena of living organisms by physical methods; the study of physical phenomena exhibited by living organisms or parts.

Biosphere

The lower part of the atmosphere, upper part of the soil layers and top portion of seas wherein all living organisms exist.

The Earth along with its living organisms and atmosphere which sustains life. Here the atmosphere means the air, land and water in and on which different beings exist.

Biotic environment

This is also known as living environment. In this, an organism comes into regular contact with other organisms of its own species or different species. In this environment, the effects of pests and diseases and their causative organisms are considered, to study the performance of a crop.

Biotic factors

Living components of environment that influence growth and development of plant which include micro-organisms, plants of the same or other species and animals including man.

Black body radiation

The radiation 'radiated' by an ideal black body. In this term the proper word is only 'radiated', rather than 'emitted', which had been used for several years. The reason is that, the importance is given to 'energy' radiated in the wave band for each wavelength interval, but not the quantity of 'emitted' radiation, for the whole spectrum. Also see 'Black body'.

Black body

An ideal, hypothetical body which absorbs all the electromagnetic radiation falling on it. It neither reflects nor transmits any radiation striking it. However, when heated, it emits all the possible wavelengths of solar radiation and becomes a perfect radiator. So, an ideal black body is a perfect absorber and a perfect radiator.

Black Cotton soil
Soils with high clay content and low organic matter deep soils (60-100 cm depth) with uniform black colour throughout the profile with unique characters swelling or shrinkage during wet and dry conditions.

Black ice
Ice on water that appears dark in colour because of its transparency.

Blind cultivation
Stirring the soil in between crop rows with animals operated implements to control weeds before the emergence of crop.

Blizzard
A weather abnormally characterized by low temperature, strong winds embedded with powdery snow or ice. In some cases, a storm may cause falling of great amounts of snow or even the snow may be picked up from the ground. The wind speed generally associated with these storms are 15 metres per second or more during which the visibility is reduced to 150 metres or even less.

Blocking
A situation which arises out of the presence of anticyclones, over a large area, the result of which is relatively stationary air masses for a few weeks.

Blowing dust
The dust particles picked up locally from the surface and blown about in clouds or sheets. This is a common feature during summer months in tropical and sub-tropical areas.

Blowing sand
Same as blowing dust. But, this term is limited to dust that is collected locally.

Blowing spray
The water particles picked up by the wind from the surface of a large body of water. For example, from a large lake, river or ocean. If water particles are picked up from oceans, they contain some elements which act as nuclei of condensation.

Bluegreen algae

An heterogenous group of procaryotic photosynthetic nitrogen fixing organisms which contain chlorophyll 'a'. They include unicellular conical and filamentous species. They are obligate phototrophs and store cyanophycean starch.

Boiling point

The temperature at which the vapour pressure of a liquid is equal to the pressure on the liquid, when in contact with each other.

Boiling

A change of state from liquid to vapour when atmospheric pressure is equal to the pressure on the liquid.

Bolometer

An instrument used to measure the intensity of radiant energy, with special reference to 'thermal' component of the wave band.

Boot stage

A stage under reproductive phase in rice where the sheath of the flag leaf (boot leaf) swells or bulges due to development of panicle.

Border irrigation

Irrigating the close growing crops by dividing the field into series of strips, each of which is flooded separately. Generally, these strips extend in the direction of the steepest slop (not more than 3%) and at right angles to the supply ditch. The strips are made level transversely but follow a gentle down field slope longitudinally. The length of strip depends upon the type of soil and crop. Otherwise called as border strip irrigation.

Border rows

The adjoining one or two rows on four sides of a plot. Biometric observations are taken excluding these rows for more precision i.e., the net plot area is considered for computation of the results.

Botanical pesticide

An important component in ecofarming or Organic farming where the pests are suppressed by the chemicals derived from plants. Also called plant-derived pesticide.

Boundary layer

This term is widely used both by meteorologists and micrometeorologists of crop plants. In micrometeorology, this is in usage to refer to a very thin layer of 'fluid' in less than or equal to a millimeter of thickness, adjacent to any 'surface' of a substance. Here, 'fluid' means either air or liquid and 'surface' means a leaf, stem, etc., of a crop in question. The usage by meteorologists is for a kilometer or less of atmosphere layer adjacent to the Earth. The common feature in both the cases is that the dynamic behaviour of the fluid should be influenced by the surface characteristics over which it exists, and which include frictional drag, porosity, undulation, etc.

Bow echo

A reader echo signature with outward "bow" shape often associated with severe thunderstorms, especially those that produce wind damage.

Bowen ratio

The ratio of vertical energy flux of sensible heat to latent heat at a given 'site' in the same direction. Here, 'site' refers to a crop canopy or across a small region over a sea or river or over an area of 2 to 4 hectares of moist land, etc. In any case, the movement of sensible heat is proportional to the temperature gradient over any interval of height. Likewise, the latent heat transfer is proportional to water vapour pressure gradient. However, both the phenomena are dominated by eddy transfer. Among others, an important condition that should exist is, that the surface should be moist in all the cases and evaporation should occur.

Boyle's law

For a gas, temperature remaining constant the product of pressure and volume is also constant, i.e., when volume increases pressure decreases.

Brand

The name, number, trade-mark or designation applied to a chemical product of any particular description by the manufacturer, distributor, importer or vender.

Bread wheat
Also known as common wheat mostly used for chapati making and most commonly grown wheat, a hexaploid of the group Triticum aestivum. About 90% of wheat area is under this species in India.

Bright line spectrum
A spectrum in which there are bright lines at specific wavelengths.

Brightness
A subjective description of light intensity. The brightness of a leaf is a reflection of its water content and physiological activity.

British thermal unit (BTU)
The quantity of heat required to increase the temperature of one pound of water by one degree Fahrenheit.

Broad bed and furrow
A successful crop production and land management of vertisols under semi arid conditions developed by ICRISAT. Raised broad beds of 90-100 cm wide and furrows of 25-30 cm deep are prepared for providing better drainage in the event of excess moisture conditions or stores water under scanty rainfall conditions.

Broad casting
A method of application of fertilizers, seeds or manures spreading uniformly all over the field by hand or by implements.

Broad-base terrace
A ridge type terrace 25-50 cm height and 5 to 10 meters wide gently sloping sides, a rounded crown and a dish shaped channel along the upper side, constructed to control erosion by diverting run off along contour at a non-scouring velocity.

Broken
The amount of sky cover for a cloud layer between 5/8ths and 7/8ths, This is done based on the summation layer amount for the layer.

Brownian motion
The irregular, short period, microscopic movements of colloidal particles in a fluid. This is also known as Brownian movement.

Bubble high

A small "high" created by precipitation and vertical instability associated with thunderstorm activity. A product of downdrafts, it is relatively cold and often has the characteristics of a different air mass. Convergence along the leading edge of a bubble high may help from additional thunderstorms. Related term : meso high

Bulk density

It is the ratio of oven dry weight of the soil to its bulk volume. It is expressed grams/cubic centimeter (c.c)

Bulk fertiliser

The solid or liquid fetilizers delivered to purchaser in non-packed form.

Bulk sample

Sample received in the seed laboratory for testing its quality parameters.

Bulk volume

The volume of an arbitrary soil mass including the volume of the solid particles and of the pores (interstices, voids).

Bulky organic manures

Huge quantities of manures that supply plant nutrients in smaller quantities and organic matter in larger quantities.

Bund former

An agricultural implement drawn with animals or tractor to make the ridges or bunds or irrigation channels to hold water.

Bund

An artificial earthern embankment made across sloping agricultural land to cut short lengthy soil slopes and reduce run-off and erosion.

Buoyancy

The 'lift' experienced by a body by virtue of which it floats in another medium because of the density difference between the substance. Here, the word 'lift' means the net force exerted upward.

Burst of monsoon

In any area a sudden arrival or intensification of air masses associated with a summer season.

Bush fallowing

A fallow system in which bushes are allowed to propagate themselves.

Buys Ballot's Law

Describes the relationship of the horizontal wind direction to the pressure distribution. In the Northern Hemisphere, if one stands with one's back to the wind, the pressure on one's left is lower than the pressure on one's right. It is reversed in the Southern Hemisphere. This law was named after the Dutch meteorologist, Buys Ballot, who developed the formula in 1857.

BWER

Acronym for Bounded Weak Echo Region. Refers to radar echo signatures with low reflectivity in the center, surrounded by higher reflectivity. It is usually associated with strong updrafts and is found in the inflow region of a thunderstorm, on a sunny afternoon.

C

C.G.S. system

Centimetre – gram – second system. A metric system of measurement in which the physical units are derived from these basic units. This system is superseded by S.I. units, for all scientific purposes.

C_4 plant

A plant which produces oxaloacetic acid containing four carbon atmos in the beginning of photosynthetic process. These plants are considered as photosynthetically efficient, because they do not exhibit photorespiration. The C_4 photosynthesis is also known as Hatch-Slack pathway or sugarcane-maize pathway, e.g.sugarcane, maize, etc.

C_3 plant

A plant which produces phosphoglyceric acid containing three carbon atmos in the beginning of photosynthetic process. These plants are considered as photosynthetically inefficient because of the presence of photo respiration, e.g., wheat, rice, barley, etc.

Calcareous soil

Soil containing sufficient calcium carbonate along with some magnesium carbonate when treated with cold 0.1 N hydrochloric acid.

Calcium ammonium nitrate (CAN)

A commercial nitrogenous fertiliser, consisting of ammonium nitrate and powdered limestone or dolomite, containing 20.0 per cent nitrogen. One half of the nitrogen is in the nitrate form and the remaining half in the ammoniacal form.

Calcium cyanamide ($CaCN_2$)

A commercial product consisting principally of calcium cyanamide and carbon. It contains not less than 20 per cent nitrogen which is not immediately plant-available.

Calcium metaphosphate ($Ca(Po_3)_2$)

The product obtained by treating phosphate rock with gaseous phosphorous pentoxide (P_2O_5) at high temperature.

Calcium nitrate ($Ca(NO_3)_2$)

The calcium salt of nitric acid. It is an excellent source of the nitrate form of nitrogen and of water-soluble calcium. The commercial product contains about 15 per cent nitrogen and 28 per cent CaO.

Calender of operations

A graphic or tabular presentation showing all kinds of farm operations to be performed during the season and specified time limits which work must be done.

Calm

Wind with a speed less than 2 nautical miles per hour or zero on Beufort wind scale.

Calorimeter

An instrument used to measure the quantity of heat and occasionally the solar radiation also.

CAM photosynthesis

Crassulacean Acid Metabolism photosynthesis which is found in many succulent plants. CAM plants close their stomata during the day and reduce losses of water by transpiration. Due to the uptake of CO_2 the acid content of these plants increases at night and decreases by day. The diurnal fluctuation is mainly due to changes in malic acid, e.g., Bryophyllum , Cactus, etc.

Cane sugar
A non-reducing sugar (sucrose) present in sugarcane, sugar beet, ripe fruits etc.

Canopy architecture
This is defined as the size and shape of canopy elements as well as their distribution in space and time which determine the physical characteristics of the canopy structure.If leaf orientation is horizontal it is called planophyll canopy. If largely vertical then it is erectophyll, and largely oblique canopies are called plagiophyll. The extremophyll canopies posses both vertical and horizontal leaves.

Canopy
The uppermost layer consisting of crowns of trees or shrubs in a forest. In a crop field, the uppermost layer of crop spread.

Canvas dam
A water control device consisting a triangular piece of canvas cloth fitted with a bamboo stick on one side and used for obstructing flow of water in channels.

Capacity factor (channel)
Ratio between actual supply to the capacity of a channel or a canal.

Capillary porosity
Percentage volume of porespace occupied by the capillary water against the gravity. Moisture storage of a soil is function of capillary porosity.

Capillary action
The behaviour of a liquid whose surface is in contact with solid such as a tube or other narrow surface due to unbalanced intermolecular attraction at the liquid boundary layer. Also known as capillarity.

Capillary conductivity
Under unsaturated conditions it is the distance travelled by soil moisture under unit potential gradient in a unit time.

Capillary fringe
Vertical distance above a water table along which moisture content varies in between saturation to field capacity.

Capillary potential
The attractive forces with which water is held by soil. It is usually expressed in terms of work that must be done to move water against the capillary forces of the soil.

Capillary rise
The rise of water above the hydrostatic surface through the influence of capillarity that is governed by transpiration pull.

Capillary water
- This is the water which is present in the capillary interspaces of soil particles as it is held by capillary forces of these soil particles. This is of utmost importance to plant life.
- Water held by micro pores of the soil or moisture held by the surface tension forces as a continuous film that is available to plants.

Capillary
In soils, very fine continuous pores through which water can pass. Because of the fineness of the pores, water can be drawn up against gravity, by capillary action thus coming into the rooting zone of the plants. In cropped fields, intercultivation and other management practices are carried to ameliorate the capillary pore spaces, so that the crop respond well to all the inputs supplied.

Captive balloon
A buoyant balloon tied to a point on the ground so that it will not move like a free balloon.

Carbon dioxide compensation point
The equillibrium value reached by a leaf when placed in a closed chamber and illuminated. This situation occurs because of the increase in carbon dioxide concentration.

Carbon dioxide fertilization
A hypothetical theory which states that, the increase in the carbon dioxide content in the atmosphere may result in the increase of the efficiency of the photosynthetic process of a crop because of its optimum growth and development when water supply is unlimited from the soil.

Carbonnitrogen ratio (C:N ratio)

The ratio by weight of organic carbon to total nitrogen in a soil or in organic material. The ideal ratio should be 10:1 to 12:1 if it is more than that the availability of nitrogen is temporarily locked up.

Cardinal points

The four chief directions of the compass i.e., North, South, East and West. These are used mainly to indicate the wind direction in met observatories.

Cardinal temperature

The minimum, optimum and maximum temperatures for the growth of an organ or whole of a crop plant. The cardinal temperature for germination in wheat is $3 - 4.5$ °C minimum, 25 °C optimum and maximum is 30-32° C. Below or above these temperature range is detrimental for plant growth and development.

Carotenoids

Yellow, orange or red pigments which resemble carotenes. Some can act as precursors of vitamin A.

Carry over soil moisture

The residual moisture available in soil that carried over to succeeding crops.

Cash-crop

A crop that fetches high marketable price for its produce. e.g. sugarcane, cotton, jute, tobacco, tea etc.

Catch crop

A short duration crop planted and harvested between a crop sequence to effectively utilize residual moisture and nutrients, usually a legume crop.

Catchment area

An area from where entire run off collected and then to a reservoir. Otherwise known as drainage basin or water shed.

Catchment basin
A unit watershed area from which all the drainage water passes into one stream or other body of water.

Cathode rays
A stream of electrons emitted from a highly charged negative body.

Ceiling balloon
A balloon used to determine the height of the cloud base above the point of observation.

Ceiling leaf area index
Leaf area index at which net photosynthesis is zero i.e. gross photosynthesis equals the respiration losses of the whole crop as result of mutual shading of leaves.

Ceiling
In meteorology and climatology, this is the height between the base of the lowest layer of the cloud and the point of observation when sky cover is atleast three fifths of the total visible area.

Ceilometer
An automatic instrument used to measure the height of the base of a cloud above the point of observation.

Celestial equator
A great circle in which the plane of the Earth's equator, meets the celestial sphere.

Celsius scale
A centigrade temperature scale, in which the melting point of ice is zero degrees and the boiling point of water is one hundred degrees. This scale is superseded by the International Practical Temperature Scale of 1948 which is expressed both in kelvins and degrees Celsius. One degree kelvin is equal to Celsius temperature plus 273.16.

Celsius temperature scale
See 'Centigrade temperature scale'.

Centigrade temperature scale

This is a temperature scale officially abandoned by International agreement in 1948, but is still in effective usage. This is the initial name for Celsius temperature scale in which the melting point of ice is zero and boiling point of water is hundred degrees respectively, at standard atmospheric pressure. This is also known as Celsius temperature scale.

Centre of gravity

The fixed point in, near or at which the body can be balanced by a single upward force. In a uniform gravitational field, this is identical to the center of mass.

Cereals (crop)

It is derived from the word 'ceres' the roman goddess of harvest of grain crops. These grain crops are called as cerealia munera and subsequently as cereals. Cereals are grain crop plants belonging to the grass family grown and utilized as staple food by mankind and feed/fodder for animals.

Certified seed

Progeny of breeder, foundation or registered seed, which is produced by maintaining, prescribed standards in terms of genetic purity and physical mixture that is acceptable to certifying personnel.

Chaff

- The thrash of plant parts and debris separated from seed while threshing of a crop or processing of a grain.
- Aluminium covered nylon strips of one centremetre length with opposite electrically charged ends. These are injected into the clouds during thunder storms to reduce the intensity of damage.

Change of state

A change in physical condition of a substance from one state of matter into another. Heat changes occurs during all changes of state, e.g. condensing, freezing, melting, etc.

Changeable weather

In the sky, huge amounts of cloud cover appear and change abnormally. Large low clouds occupy and disappear occasionally under this condition.

Chaos theory

The theory which states that the climate of a region or a given location makes sudden 'jumps' resulting in unimaginable behaviour, which is contrary to the normal.

Characteristic radiation

Radiation originating from an atom following the removal of an electron. The wavelength of the emitted radiation depends only on the element concerned and the particular energy levels involved. It is to be stated that all the bodies above absolute zero temperature emit radiation.

Characteristics curve (moisture release curve)

Graphical representation of relationship between water content and water potential (soil suction, matrics suction) of a soil.

Charles law

The volume of a given mass of gas is directly proportional to the absolute temperature at constant pressure.

Check dam

A cement and concrete structure built across the water course or in the stream to store, distribute the water besides controlling soil erosion, recharging ground water table.

Check irrigation (check basin irrigation)

A surface irrigation method for close growing crops. Where the area is devided into small plots with levees or checks to irrigate the plots at uniform depth.

Chemical fallow

Controlling of weeds on a fallow land with substitution of chemicals and tillage implements.

Chemotropism

Changes in movements and growth in response to chemical.

Chestnut soil

A soil having a dark-brown surface developed under mixed tall and short grasses in a subhumid to semi-arid climate. The combination of tall and short grass species in subhumid to semiarid area develops the soil to a dark-brown colour

Chick pea

A winter grain legume crop cultivated during Rabi season also called as gram, bengal gram.

Chilean nitrate of soda

A product obtained by refining the crude nitrate deposit found in Chile and containing about 99 per cent sodium nitrate also called as chile salt petre.

Chinook

A warm and dry wind descending the eastern slopes of rocky mountains in North America onto the adjacent plains. This is categorized as local wind.

Chisel

An implement to till the soil up to a depth of 20-40 cm to break the hard soil layers.

Chloroplast

A plastid in which photosynthesis is carried out. The plastids exist in all photosynthetic organisms except photosynthetic bacteria and blue green algae. In advanced plant communities there may be a hundred or more chloroplasts in a single cell.

Chromosphere

The layer of the Sun's atmosphere outside the reversing layer. This is visible during a total eclipse and is several thousand kilometers deep and has an estimated temperature of twenty thousands of degree Kelvin.

Cirrocumulus

A high level cloud and the mean height is 5 to 13 kilometres. The prefix 'cirro' refers to 'hair band like' or 'hair lock'. These clouds are in the form of heaped hair locks, white in colour, transparent in nature and produce halo phenomena without shading. Also, look like thin white parches or sheets.

Cirro-stratus

A high level cloud and the mean height is 5 to 13 kilometres. The prefix 'cirro' refers to 'hair band like' or 'hair lock'. Transparent, whitish cloud veil of fibrous or smooth appearance. The prefix 'stratus' refers to layers and the clouds form a very thin layer which gives the sky a milky appearance.

Cirrus

A high level cloud and the mean height is 5 to 13 kilometres. The prefix 'cirro' refers to 'hair band like' or 'hair lock'. Detached clouds in the form of white delicate filaments. At sunrise and sunset, these clouds give marvelous colours and do not give precipitation.

Clausius clapeyson's equation

The equation which reads that, for a liquid the boiling and freezing points depend on pressure.

Clay atmometer

An instrument used to determine the evaporation.

Clay soil

A soil that contains a minimum guarantee 40% of clay but sand, silt are less than 40%.

Clay pan

Otherwise called as plow pans formed under puddling conditions in sub soil, hard enough that is impediment for root growth.

Clay spiles

A water control device made up of Chinese clay with a diameter and length 8 cm and 30 cm respectively used for division of water.

Clayey

All clay textural classes viz, sandy clay, silt clay and clay under one umbrella.

Clean cultivation

In a growing crop all vegetation and thrash is eliminated by all possible means.

Cleaning

The removal of foreign or dissimilar material by washing, screening, hand picking, aspiration or any other mechanical means from seed or seed material.

Clear sky

The sky is said to be clear when the existing cover of the clouds is less than one okta

Climate factors

Atmospheric or meteorological conditions which collectively make up the climate. Even though climatic and weather factors are difficult to separate, the climatic factors include the major weather elements such as solar radiation, temperature, wind, etc.

Climate

This is derived from the Greek word 'klima' meaning inclination or slope. Climate is the generalized weather or summation of weather conditions experienced in any location or over a given region during a comparatively longer period. This depends upon latitude, altitude and contenentality. The aspects involved are 1. larger areas like a zone, state, country, a part of continent, etc., 2. longer duration of time like a month, season, an year, etc; and 3. described by average values or normals.

Climatic analogues

Area that are enough alike with respect to some of the major weather characteristics affecting the production of crops.

Climatic element

Any one of the properties or conditions of the atmosphere to indicate the physical state of the weather or climate of a given place at any moment.

Climatic formation

A complex of communities which are geographically linked with one another because of climatic conditions. An extremely complex vegetation unit is formed in an area because of the variations of climate for a given longer period, under study.

Climatic normal

The climatic normals are the degrees of temperature, amount of rainfall, humidity, etc., with distinguish optimal conditions from those defined as abnormal, both because of excess and insufficiency.

Climatic regime

A zone in which the homogenous climatic conditions exists.

Climatic region

One of the main portions of the surface of the Earth deleneated on the basis of climate such as tropical, polar, etc. The world climatic regions are classified based on some specific parameters by different scientists over a period of time.

Climatological observation

Recording, analysis and interpretation of data of one or more weather and climatic elements to arrive at their usefulness for several purposes.

Climatological station

A station at which the climatic elements such as maximum and minimum temperature, sky condition, type of precipitation, number of sunshine hours, direction and velocity of wind and other weather data is scientifically observed, recorded and maintained as per the IMD guidelines, in India.

Climatology

Derived from Greek language. 'Klima' means slope of the Earth and 'logos' means study. This can be stated in the following ways:

The science dealing with the factors which determine and control the distribution of climate over the Earth's surface. The science which studies the totality of weather.

Climax community

A self perpetuating community that is stable and in complete harmony with physical and biotic environment where it adapted.

Climax vegetation

The most fully developed natural plant community sustained by climate.

Climograph

A chart in which one climate factor such as the mean monthly temperature is plotted against another factor such as the mean monthly precipitation or the mean relative humidity.

Climosequence

A series of climatic data for different places in an area. Such compilation is done to find the integrated effect of weather elements on crop production.

Clod

A coherent mass of soil ranging in size from 5 to 10 mm to as much as 20 to 250 mm produced artificially, usually by the activity of man in ploughing, digging etc. especially when these operations are performed on soils that are either too wet or dry for normal tillage operations.

Clothes line effect

Movement of advective wind from a warm up area to a comparatively cooler crop field.

Cloud base

In a layer of cloud, the lowest level observed in the atmosphere at which the air contains precipitable quantity of cloud particles. This can be determined by ceilometer, ceiling balloon, etc.

Cloud burst

A situation in which sudden and heavy rain occurs.
A sudden and extremely heavy downpour of rain.

Cloud classification

The clouds are classified according to their shape, colour, height, process of formation, etc. The cloud classification is a scheme by the above characteristics which helps in distinguishing and grouping the clouds. The recording of amount of cloudiness is mandatory in most of the meteorological observatories.

Cloud height

The height between the point of observation and the base of a cloud.

Cloud seeding

A process in which the precipitation is encouraged by injecting artificial condensation nuclei. Aircrafts are used to spray salt nuclei, Ammonium nitrate, solid carbon dioxide, etc., to lower the temperature of the turbulent cloud. The rain starts falling within minutes after the initiation of the process. The placing of materials such as silver iodide in the clouds to produce precipitation.

Cloud types

(a) High clouds (b) Medium clouds (c) Low clouds (d) Clouds with vertical development.

Cloud

An aggregation of minute drops of water and or ice crystals present in the air in an indefinite unit with its base above the ground. The clouds form when rising air cools to such an extent that it can no longer hold all the water within it as water vapour.

Coarse texture

Includes the sands, loamy sands, and sandy loams, except the very fine sandy loam textural classes. Sometimes subdivided into sandy and moderately coarse textured.

Coated fertilizer

Fertilizer, the granules of which are covered with a thin layer of a different material in order to improve the behaviour and or modify the characteristics of the fertilizer. Usually they release the nutrients slowly ex: Gypsum coated urea.

Coefficient of eddy diffusivity

The appropriate transport constant for continuous downward flux of momentum by turbulence.

Coefficient of expansion

A constant, under given conditions, which shows the linear of volumetric expansion of one unit of a given material for one degree rise in temperature. A coefficient is a factor used to measure a specified property of a given material.

Coefficient of thermal conductivity

The quantity of heat that flow in a second through a cubic centimetre of matter in which a temperature difference of one degree Celsius is maintained between the opposite faces.

Cold air drainage

The setting of cold air in low places displacing the less dense warm air. In hilly and valley regions this is observed.

Cold air

If the temperature of air is relatively low as compared to an underlying surface or another adjacent air column or within itself, then the air is said to be cold air.

Cold desert

Land covered with snow and ice. In polar regions this situation is seen.

Cold front

A 'front' advancing so that the cold air mass replaces the warm air mass. In general, upon the passage of a cold front different weather conditions exist.

The boundary of a mass of cold air and a mass of warm air.

Cold hardiness (resistance)

The capacity of a plant species to tolerate low temperature. In temperate regions, unless a particular crop species possess the cold hardiness, it will not survive and produce desired yields.

Colloidal clay

The smallest unit of clay usually $\leq$ one micron diameter that controls most of the physical and bio-chemical properties of the soil.

Columnar soil structure

Similar to prismatic structure except that the tops of the blocks are rounded.

Combination method

A method by which the evapotranspiration is estimated using the climatic elements as input.

Combined tillage

A time saving tillage system in which two or more tillage implements are combined to reduce the number of operations. This assumes very significance under dryland conditions.

Combineharvester – thresher (combiner)

A machine designed for harvesting, threshing, separating, cleaning and collecting grain while moving through the standing crop in a single stretch.

Command area

Irrigated area under an irrigation project, dam or by a canal.

Common name

Common name means the name assigned to a pesticide active ingredient by the International Standards Organisation or adopted by national standards authorities to be used as a generic or non proprietary name for that particular active ingredient only.

Community equilibrium

The condition in which a community is maintained with only minor fluctuations in its composition within a certain period of time.

Community forestry

Establishment, maintenance of tree component for avenue planting, in amusement parks by the local community or village panchayat. The end products are utilized for the community.

Companion crop

Any subsidiary crop grown in association with a main crop.

Compensation intensity

The intensity of light at which the amount of oxygen produced by photosynthesis of a plant equals the oxygen absorbed in respiration, while the opposite is true for carbon dioxide released in respiration. This is also known as light compensation point.

Complementary colours

Two spectral colours which add together to make white or grey.

Complementary intensive cropping

Growing of morphologically and physiologically different crops in an agricultural year under the sequential intercropping systems as to produce complementary effects upon each other and on succeeding association.

Complete fertilizer

A product obtained by mixing different fertilizer stock materials, containing three major plant nutrients, namely, nitrogen, phosphorus and potassium (See also Fertilizer Mixture)

Complex fertilisers

The commercial fertilisers containing at least two or more of the major nutrients, made by a chemical reaction between the nutrient containing raw materials.

Component crops

Individual crops making up a cropping system.

Compost

Reinforcing a bulky organic manure that obtained by subjecting the farm and home wastes to controlled decomposition.

Concentrates (feed)

Feed that contains all necessary ingredients to form a complete and balanced ratio for dairy, piggery, poultry, fish etc.

Condensation nucleus

See 'nuclei of condensation'.

Condensation trail

A cloud like streamer frequently observed to form behind aircraft flying in clear, cold humid air.

Conditional instability

When a parcel of air is rising in the atmosphere its temperature profile changes according to the wetness or dryness of that parcel. The region in which these changes occur is said to be in conditional instability when a wet parcel is unstable and a dry parcel is stable. In meteorology, this situation can be seen when a previously unsaturated layer of air mass is filled with a cloud or clouds. This phenomenon is also called as convective instability.

Conduction

A method of transfer of heat through matter from places of higher to places of lower temperature, by the interaction of atoms or molecules possessing greater kinetic energy with those possessing less.

Conductive capacity

The product of the heat capacity and square root of the thermal diffusivity.

Conductivity of gases

In gases, the heat energy is transmitted by collision of the gas molecules, and determination of this conductivity is very difficult because of the radiation and convection. Since the convection is negligible in gases, the difference between the total loss of energy and the loss due to radiation is equal to the loss of heat due to conduction.

Conductivity of liquids

Like conductivity of gases, in liquids also the heat energy is transmitted by collision of molecules. If it is gas molecules in gas, it is liquid molecules in liquid that take part in the transfer of heat.

Conductivity

The ability to transfer the energy through the atoms or molecules of the conductor.

Conservation of charge

The word 'charge' is applied to 'electric charge' but there is no complete definition for this term as on today. In defining the properties of matter, this is being satisfactorily used. The law of conservation of charge states that, in a closed system the total electric charge does not change.

Conservation of energy

The capacity for doing work is energy. Heat, kinetic, electrical and radiant energy etc., are different forms of energy which are interconvertible by suitable means, but it can occur only in the presence of matter. The law of conservation of energy state that in a closed system the total energy does not change.

Conservation of momentum

The product of mass and velocity of a body is momentum. The law of conversation of momentum states that, in a closed system the momentum does not change. So, as per this principle, the total momentum of two colliding bodies before impact is equal to their total momentum after impact.

Conservation tillage

A tillage system discovered to maintain roughness of a field surface and at least 30% of surface is in roughness, reduces run off water and wind erosion.

Conservation

The maintenance of the environment in a socially acceptable form.

Constant level balloon

A balloon which floats at a constant pressure level in the atmosphere.

Consumptive use

Quantity of water irrespective of source that is lossed through evapotranspiration by crop plants as well as that part of water utilized in plant metabolic activities. Since water used in plant metabolic system is negligible ($< 1\%$ of total E_T) it is counted on par with water use efficiency.

Contact anemometer

An electrical instrument used to measure the speed of the wind.

Contact herbicide

In this the mode of action of a chemical is through physical contact rather than translocation.

Contact placement (fertiliser)

Sowing of seed and fertilizer combindely with help of seed drill, seed cum fertilizer drill.

Continental climate

The climate of the continent particularly in its interiors. There is a large variation in annual and daily range of temperatures. For instance, humid continental climates are warmer but less severe.

A type of climate which is experienced in large land masses located in interior parts of the continents. The characteristic feature of the climate is that a large seasonal range of temperatures, as the oceans are located substantially away from these areas,

Contingency cropping

Under aberrant weather conditions or pest epidemics growing of substitute crops aiming at partial mitigation of misery of producing some food, feed and fodder to encounter emergency conditions.

Continuity equation

Formal description of the conservation of the mass of a body of moving air relating to net mass flux into a fixed volume to the rate of accumulation of air mass there.

Continuous grazing

Feeding of a particular range or pasture by livestock round the year or grazing season.

Contour border irrigation

A method of surface irrigation in which channels are laid out in strips along the contours across the slope.

Contour bunding

The construction of small bund across slope of the land on a contour so that the long slope is cut into a series of small ones and each contour bund acts as a barrier to the flow of water.

Contour cropping

A method of practice in which crops are cultivated along the contour lines across the slope.

Contour farming

A soil moisture conservation practice of rainfed areas with undulating topography where all agricultural, allied operations are done along the contour across the slope.

Contour interval

The differences in height of two adjacent contour lines on a contour map.

Contour irrigation

A method of surface irrigation where water is applied to field along the contours across slope.

Contour planting

A method of sowing in which trees or crop rows are planted across the slope along the contours.

Contour ploughing

A cultivation practice in which mechanical manipulation is done along the contours across the slope.

Contour strip cropping

An agronomical method of soil conservation practice involving growing of a soil exposing and erosion permitting crop in strips of suitable widths alternating with strips of soil protecting and erosion resisting crops, along the contours across the slope.

Contour terrace

A mechanical method of soil conservation practice in which low terrace formed along the contours across the steep mountains slopes with inadequate natural cover to prevent erosion until natural cover becomes established.

Contour

- Means an imaginary points of the similar or same height on the earth surface.
- A curve on a drawing chart or a synoptic chart, etc., connecting points of equal elevation. In meteorology and climatology, this is a height curve of an isobaric surface.

Control plot

A standard check plot used for comparing the relative performance of treatments under study.

Convection current

The process of vertical transport of electric charge by the mass motion of a charged medium. This term is used mainly for the vertical motion of the air itself but may be considered to include the 'precipitation current' associated with the fall of charged precipitation particles. Sometimes, the movement of air caused by heat and temperature differences is also called by this term.

Convection

The transfer of heat through a liquid or gas by the actual movement of the liquid or gas molecules caused by temperature difference. In climatology, the term is used for the upward movement of relatively warm air. In fluids, the transfer of heat by current is referred to as convection. In such cases, convection results in mixing of the properties of that fluid.

Convective condensation level (CCL)

The lowest level at which condensation will occur as a result of convection due to surface heating. When condensation occurs at this level, the layer between the surface and the CCL will be thoroughly mixed, temperature lapse rate will be dry adiabatic, and mixing ratio will be constant. This situation, usually occurs in summers in tropical and sub-tropical areas.The state of an unsaturated parcel or layer of air which produces instability if it is lifted to attain saturation is called potential or convective instability The level above which such a lifted parcel of air becomes freely convective because of condensation is called as convective condensation level. This happens during the heating of underlying surface.

Convective instability

The state of an unsaturated layer of air whose lapse rate of temperature and moisture are such that when lifted adiabatically until the layer becomes saturated. Usually, convection is spontaneous. When the formation of cumulonimbus cloud takes place this phenomenon can be observed.

Conventional tillage

A customary way of land preparation for sowing that includes primary tillage followed by secondary tillage operation.

Convergence

Coming to a point. The negative of divergence.

Cooperative better farming

A type of farming where the land is not pooled and the cultivation is carried on by each farmer separately. A member is free to form his own way of farming except in respect of the purpose for which he has joined the society e.g. for irrigation, purchase of seed or marketing of produce etc.

Cooperative collective farming

A type of farming where in the land is owned by the society and cultivation is carried out jointly. The members work on the land under the direction of a managing committee. The profits are paid to the members in proportion to the work and capital contributed by each member. The right or share of individual member in the land is not recognized.

Cooperative farming

A system under which all agricultural operations or part of them are carried out jointly by the farmers on a voluntary basis; each farmer has retaining right in his own land. The farmers pool their land, labour and capital. The land is treated as one unit and cultivated jointly under the direction of an elected management. A part of profit is distributed in proportion to the land contributed by each farmer and the rest of the profit is distributed in proportion to the wages earned by each farmer.

Coperative joint farming

A type of farming, in which land of members is pooled for joint cultivation. The ownership of each member over his own land is recognised by payment of a dividend in proportion to the value of land. The members work under the direction of the managing committee and each member receives for his daily labour.

Cooperative tenant farming

A type of cooperative farming wherein society is owner of the land. The land is then divided into plots which are leased out to members. The society arranges for agricultural requirements e.g. credit, seeds, manures, marketing of the produce etc. Each member is responsible to the society for payment of the rent of his joint plots.

Coppice farming

The word 'coppice' has been defined as the ability to regenerate by shoot or root suckers; thus coppice farming is a practice of intensive exploitation of land capability by regular harvesting of such trees which produce numerous shoots from the stump after the cutting. Coppice sprouts usually grow vigorously because they are served by roots big enough to feed the former tree.

Coriolis acceleration

The acceleration arising from the motion of a body corresponding to the rotating terrestrial reference frame.

Coriolis effect

Because of the rotation of the Earth the deflection of wind in the atmosphere takes place in a rhythmic way from a straight imaginary path. The deflection is to be right in the northern hemisphere and to the left in the southern hemisphere, because of the coriolis force effect. Also see 'Coriolis acceleration'.

Coriolis parameter

The vorticity of the Earth about the local vertical, in which the Earth's angular velocity and angle of the vorticity are related.

Corona

The outer white irregular halo of the Sun, which is visible during total eclipse

Correlation coefficient (r)

The degree of correlation is called correlation coefficient. It varies from +1 to −1. A correlation coefficient of zero means that the two variables are not interrelated. A 'r' value of +1 or -1 indicates complete association. A positive correlation means that high values of one variable are related with high values of the other. A negative correlation means that as one variable decrease the other tends to increase.

Correlation

Association between two variables such that an increase or decrease of one is linked with an increase or decrease in the other.

Cost of cultivation

All the expenditure involved in production of a crop that includes rental value also.

Countability

In differential calculus, a set is called countable or enumerable if its elements can be placed in 1-1 correspondence with the natural numbers.

Counter radiation

During the day time, when the Sun shines, the Earth gets heated up by absorbed solar radiation. In the night it loses the energy in long wavelengths and a large part of this radiation is absorbed by the atmosphere and sent back to Earth's surface as counter radiation. Because of this counter radiation the Earth doesn't become cooled completely.

Covariance

The mean of the product of corresponding deviation of two variates from their individual means. Estimated as the ratio of the sum of cross products to the corresponding number of degrees of freedom.

Cover crops

Crops which reduce the loss of moisture through evaporation and erosion through water and wind by covering the soil surface.

Creeping

This is the defect in an aneroid barometer as a result of rapid and large changes in pressure, which can be rectified by an adjustment.

Critical angle

The least possible angle of incidence at which total internal reflection occurs. This can be stated as the angle of incidence for a given material at which a ray of light entering a vacuum from the material is refracted at ninety degrees.

Critical difference

The square root of the error mean square, measures the standard error per plot due to uncontrolled environmental effects. The standard error of mean is determined by dividing standard error by number of observations entered into in calculating the mean. The standard error of difference when multiplied further by t-value (at the required level of significance) at the error degrees of freedom, is known as critical difference. The testing of significance of differences between individual treatments should not be done in a trial unless overall significant differences between treatments are indicated by 'F' test in the analysis of variance.

Critical isothermal

The isothermal relative to the critical temperature.

Critical period or critical stages (crop)

The period or the stage of development in the life cycle of a crop or a plant that is most sensitive to the deficiency of a production factor and is most responsive to the correction of the deficiency. The production factor may be water, nutrient or free from weeds, this stage varies according to crop.

Critical photoperiod

It is the day length value separating the inductive range of photoperiod from non-inductive range, for example, the critical photoperiod for flowering in Xanthium is 15 hrs.

Critical pressure

The pressure of the saturated vapour of a gas (substance) at its critical temperature.

Critical region

That portion of the area under curve which includes those values of a statistic which lead to rejection of the null hypothesis.

Critical temperature

This is the temperature at which a liquid vapourises regardless of pressure. This can also be said as the highest temperature at which a gas can be liquified by the increase of pressure only.

Critical value

A value below which the nutrient under study is inadequate to meet the requirement.

Critical volume

The volume occupied by a unit mass of a substance at its critical temperature and pressure.

Crop cafeteria

Crop cafeteria is the demonstration of identified efficient crops/varieties in an agrometeorological region/zone offering an opportunity to the farmer to choose a suitable crop or crop combination commensurating with the available resources and requirements. This approach is also applicable in case of production technology where a number of them are demonstrated and the farmer has a choice to choose a suitable one. Maintanance of crop cafetaria is also a mandate of college farm for student instructional purpose.

Crop calendar

A list of the standard crops of a region in the form of a calendar giving the dates of sowing and various operations including harvesting during the crop seasons in years of normal weather.

Crop canopy

It is the structure of aerial vegetative parts with special reference to the size, orientation, depth, density, and arrangement of leaves influencing the penetration and interception of solar radiation.

Crop coefficient

The coefficient is a factor or multiplier which measures some specified property of a given substance. The crop coefficient is a broader term used in the evapotranspiration calculations, and it is the ratio of evapotranspiration occurring with a specific crop at a particular stage of crop growth to a reference crop evapotranspiration at that time. This is the ratio between the potential evapotranspiration of a crop and the potential evapotranspiration of the grass. The crop coefficient is assumed to be a constant depending upon ground cover i.e., the stage of the crop or plant and prevailing meteorological conditions. In micrometeorological studies PET is frequently calculated for a grass cover of relatively smaller area and uniform height, but these values can be converted to any cropped surface by the crop coefficient.

Crop competition

Two plants are in competition with each other when the growth of either one or both of them is reduced or their form modified as compared with their growth or form in isolation.

Crop ecology

A branch of 'plant ecology' that deals with the study of the inter-relation amongst crop plants and environment.

Crop ecosystems

Cropping systems ranging between monoculture and multi-species culture of field and garden crops, singly or in combination and their relation to environmental conditions and management practices.

Crop efficiency zone

A zone or a geographical area characterised by high spread of a crop combined with high yields consistent with minimum variation from year to year.

Crop equivalent yield (cE$_y$)

It is the conversion of crop yields into one form to compare the crops grown under mixed/inter-cropping/System.
Conversion is done into monetary value/protein/carbohydrate/intrinsic energy value. The formula is $Y_1 \times P_1 = Y_2 \times P_2$

Where in an intercropping system

Y_1 = Yield (Q/ha) of base crop,

P_1 = Price of basecrop (Minimum support Price)

Y_2 = Yield (Q/ha) of inter crop,

P_2 = Price of inter crop thus.

Crop growth model

A simplified representation of a 'system' is 'model. Here, 'system' is referred to as a limited part of reality that contains interrelated elements. It is a fact that the crop growth is the integrated result of a number of interrelated physical, physiological and ecological processes that occur during its growing period. So, crop growth modeling is an art and science for simplified representation of dynamic crop production system, which has enough potential for yield forecasting and crop management. Also see 'Simulation, Mechanistic, Mathematical, Static, Empirical, Numerical, Dynamic, and Analytical models'.

Crop intensification

The concept, approach, method and process of growing more crops per year by increasing cropping intensity.

Crop logging

A foliage analysis of assessing nutrient status and water content of selective leaves at different crop growth stages. The aim is to maintain optimum nutrient or moisture content at least during initial stages for better yields.

Crop physiology

The study of interaction of integrated processes influencing the crop yield.

Crop potential

The ability of a particular crop or genotype to yield.

Crop production

It is an art and science of exploiting plant morphological and physiological characters aiming at maximizing photosynthesis with an objective of increasing the economic crop yield.

Crop productivity

The crop yield per unit area or the phytomass of the crop per unit area. The yield may be of grain, stem or biomass. The vertical increase in yields of crops is the latest methodology being followed in a country like India. This helps in better fertilizer, water and nutrient management, in addition to better land utilization. It is expressed as kg/hectare or tonnes/hectare.

Crop residue management

Use of the non-commercial portion of the plant or crop for protection or improvement of the soil.

Crop residue

The portion of a plant or crop left in the field after harvest, threshing or that part of the crop that is not used domestically or sold commercially.

Crop rotation (sequential cropping)

The growing of different crops on a piece of land in an ordered succession.

Crop technology

The theoretical and practical approach to crop production in which concepts of applied science and advanced techniques are related to the financial consequences of their use.

Crop water use efficiency

The ratio of crop yield to the amount of water depleted by the crop plants through evapotranspiration. It is expressed as Kg/ha-cm

Crop yield index

A measure of comparison of the yields of all crops on a given farm with the average yield of these crops in the locality. The relationship is expressed in percentage.

Crop

A group of plants, or aggregation of individual or different species grown over a unit area for a specific purpose in which an economic criteria is involved.

Cropping index

The number of crops grown per annum on a given area of land times 100.

Cropping intensity

Is the ratio between total cropped area and actual net cultivated area expressed in percentage.

Cropping pattern design

The crop configuration or sequencing done on paper for year-round land utilization at a given area considering physical, biological, and socio-economic factors prevailing at that area.

Cropping pattern testing

The growing of a designed cropping pattern at a given site and evaluating biological stability, agronomic productivity, and economic profitability.

Cropping pattern zones

It is developed to divide the country into homogeneous units using the entities like soil and climate besides physical and agronomic criteria subdivided on the basis of isothermic lines.

Cropping pattern

Yearly sequence and spatial arrangements of crops and fallow on a given area. Arrangement of crops in space and time dimensions on a farm or group of farms in a farming year.

Crop Resistance Index (CRI) :

It is one of the indicators of weed suppression by imposing a weed control treatment in a crop or cropping system. Higher the value higher will be the grain yield and is given by the formula,

$$CRI = \frac{\text{Dry matter production by crop in the treatment plot}}{\text{Dry matter production by crop in the control plot}} \times \frac{\text{Dry matter production of weed in control plot}}{\text{Dry matter production of weed in treated plot}}$$

Cropping scheme

The plan according to which crops are raised on individual plots of a farm with an object of getting the maximum returns from each crop without impairing fertility of the soil.

Cropping system

Cropping pattern on a farm or in a region or order in which the crops are cultivated on a piece of land over a fixed period and their interaction with farm resources and other farm enterprises and available technology which decides their proportion.

Cropping systems research

The research activities, mainly farmers' field that focus on the understanding of farmer's existing cropping systems; design, testing, and development of new improved cropping patterns and component technologies for selected environments to efficiently utilize available farm resources.

Cross drill (cross sowing)

To drill the seed in two directions usually at right angles to each other.

Crown cover

The canopy formed by the trees in a forest.

Crust

A hard or brittle layer formed on the surface of many soils when dry that affect the germination of the crop

Cryopedometer

An instrument used to measure the depth to which the soil is frozen.

Cultivable or culturable command area

Gross command area minus uncultivated area. Command area is that which falls within irrigable area of a project.

Cultivar

Is equivalent "variety" and is defined as an assemblage of cultivated plants which is clearly distinguishable by any character (morphological, physiological, cytological, chemical or others) and when reproduced (sexually or asexually) retains its distinguishing characters. The term is derived from cultivated variety.

Cultivation

Includes any labour and care taken in the raising of plants, such as stirring of soil, fertilizing, etc. or to loosen the soil around a plant for the primary purpose of weed control.

Cultural control (pests)

Control of pests by cultural practices such as soil tillage, management of fertility, weeds, water and crop residue, plant spacings and cropping pattern.

Cultural practices

Crop-care practices including land preparation, seeding or transplanting, weed control, fertilizer and insecticide application, water control in the field etc.

Cumulative run off

The total volume of run off over a specified period of time.

Cumulo congestus

A conspicuously growing cumulus cloud to a moderate or large size.

Cumulonimbus

A low level cloud and the mean height is 0-2 kilometres. These are heavy and dense clouds in the form of a mountain and associated with precipitation and lightening. These are towering clouds and at times spreading out on top to form an 'anvil head'.

Cumulus

A low level cloud and the mean height is 0-2 kilometres. The word 'cumulus' refers to 'heap'. These are detached clouds like cirrus, dense with sharp edges. The clouds develop vertically in the form of rising mounds, domes and the bulging upper parts resemble a cauliflower.

Cup anemometer

An instrument used to measure wind speed. Usually, three to four conical cups are fixed to the end of a horizontal arm projecting from a vertical axis and the speed of the rotation is observed and recorded in kilometers per second.

Curie

A basic measure of unit of radioactivity of a substance. The disintegration of a radioactive isotope at the rate of 3.7 times 10^{10} atoms of material per second. A one thousandth part of a curie is a millicurie.

Current

In physics the word is used to refer to the electric current, which means, the flow of electrons through a conductor for which the S.I. unit is ampere. In meteorology, this is used to refer the motion of a fluid in a given direction and also to the movement of air as in a convection current.

Cusec

The quantity of water flowing at the rate of one cubic foot per second. One cubic foot of water weighs 62.4 lbs or 6.24 gallons.

Cut off time (irrigation)

In surface irrigation the time at which the supply of water is cut off at the top end leaving the water already supplied to flow down the slope.

Cuticular transpiration

About 5 to 10 percent of the total transpiration from the plants which occurs through cuticles is known as cuticular transpiration. This type of transpiration depends upon the thickness of the cuticle and presence or absence of wax coating on the surface of the leaves. Xerophytic plants have very thick cuticle and wax coating on the leaves and stem, in order to check cuticular transpiration.

Cyclogenesis

Any development or strengthening of cyclonic circulation in the atmosphere.

Cyclone wave

A deformation of flow associated with an extratropical cyclone in the form of a wave.

Cyclone

An area of low atmospheric pressure into which the air flow is deflected. This deflected air rises and blows around the low pressure in an anti-clockwise direction in the northern hemisphere and the clockwise direction in southern hemisphere. In this synoptic weather system, a large storm develops in closed circulation over a thousand kilometer area. There are different names for this in different places and regions of the world, and this is the name for a severe tropical storm in the Indian Ocean.

D

Daily maximum temperature

As per the IMD guidelines, the maximum temperature recorded in a 24 hour period at any given place or location.

Daily minimum temperature

As per the IMD guidelines, the minimum temperature recorded in a 24 hour period at any given place or location.

Daily range of temperature

The range of temperature [between maximum and minimum] recorded as per IMD guidelines in 24 continuous hour.

Dalton equation

The Dalton's law of partial pressures states that the total pressure of two or more gases or vapour is equal to the sum of the pressure that each component would exert if it were present alone, and occupy the same volume as the whole mixture. In the horizontal movement of air, the Dalton's equation connect three important variables, namely, the vapour pressure of the wind, the vapour pressure of the surface and wind velocity.

Damping ratio

A decrease in the amplitude of an oscillation or wave motion with time is 'damping' . In most of the micrometeorological observations, certain factors change, all of a sudden and the instruments used also respond equally in their readings. So, the performance of the instrument in the bargain to adjust to that sudden is known as damping ratio of that instrument. For example, a wind vane responds to a sudden change in wind direction.

Damping

A decrease in the amplitude of an oscillation or a vibration or wave motion with time.

This happens in all the probability due to loss of energy.

Dapog rice nursery

Rice seedlings are raised on a concrete floor or on soil covered with banana leaves or plastic sheets fixed with bamboo pegs. Dapog seedlings are ready for transplanting in 9-14 days after the seed is sown. It is labour and time saving method with ease of transport. Otherwise it is called as soil less nursery introduced from Philippines.

Darcy's law

The law states that the rate of supply of water to the root within a given surface area is proportional to the hydraulic conductivity and also the potential gradient at the root surface.

Data Buoys

Buoys placed throughout the Gulf of Mexico and along the Atlantic and Pacific coasts of the United States. They relay information on air and water temperature, wind speed, air pressure and wave conditions via radio signals.

Data

Numbers of measurements which are collected as a result of observations.

Dawn
The first appearance of light in the eastern sky before sunrise. It marks beginning of morning twilight. The visual display is created by the scattering of light reaching the upper atmosphere prior to the sun's rise to the observer's horizon.

Day
The period of the day between sun rise and its setting (dusk and dawn).

Day cusec
In a day of 24 hours the quantity of water flowing in cusecs.

Day length
The duration from sunrise to sunset on a given day. This is expressed in hours. If the sky is completely covered with clouds during this period, it has no significance.

Day neutral plant
A plant in which flowering can occur irrespective of the length of the day, e.g., Sunflower.

Dead furrow
An indigenous practice by farmers in which a wider furrow is formed between strips of ploughed land that stores the water.

Dead storage
The storage at which the water cannot be let out in a reservoir.

Debris cloud
Considered a rotating cloud of debris or dust near the ground. The debris cloud appearing beneath a thunderstorm will most likely confirm the presence of a tornado, which may strike violently.

Decomposition
A biochemical process in which minerals as well as organic material are broken down.

Deep freezing
A technique used to store many perishable foods for human consumption by maintaining subzero temperatures. At this temperature organisms responsible for decay cannot flourish. By this technique high quality vegetables can be made available at all times of the year even though they cannot be grown due to adverse agroecological environment throughout the year.

Deep percolation loss
Vertical or down ward movement of water through soil profile out of root zone.

Deep placement
An agronomic practice in which fertilizers are placed to a depth of 20 ± 1cm before sowing of the crop. The aim is to avoid the loss of nutrients through various processes.

Deep seepage
The portion of the run off which escapes from a reservoir through the underlying earth or rock strata below any possible intercepting cutout constructed at the dam.

Deep soil
A soil having a solumn depth of more than 100 cm.

Deep tillage/subsoiling
Refers to tillage that loosens the soil 40 cm deep or more with the purpose of conserving moisture in the monsoon season or for breaking hard pans. This can be achieved with chisel plough or subsoiler.

Deep water rice
Rice grown with more than 51cm to 5-6 meter of standing water.

Deepening
In a low pressure system, the further decrease of central atmospheric pressure. It can also be applied to the same phenomenon in a high atmospheric pressure system under certain peculiar conditions and situations for explaining the system technically. But in all the likelihood, majority of the usage goes to a low atmospheric pressure system only.

Deferred grazing
Withholding animals from a pasture beyond the normal beginning of the grazing season.

Defloculate
To separate, to break down soil aggregates of clay into their individual particles.

Degree day

The departure of mean daily temperature from a reference temperature. In micrometeorlogy, for every crop and for every phenophase of each crop a reference temperature is given, the importance of which is that below that point no plant physiological activity takes place. When a cumulative figure is obtained for a particular phase, it indicates the essentiality of that plant to pass through all the time the same degree days for satisfactory growth.

Degree

A measure of temperature difference representing a single division on a temperature scale.

Dehulling

The process of removal of hulls that cover the rice grain (outer covering) from intact grains or seeds.

Delta (river)

A deposit of sand or sediment, usually triangular, formed at the mouth of rivers.

Demonstration trail

A non-replicated trail used to demonstrate some established fact or principle.

Demonstration

A pragmatic way of customery showing to the user the working of a particular practice or technology, developed and established on a research farm.

Denitrification

A biological reduction process of nitrate or nitrite to gaseous nitrogen from the soil under submerged conditions either as molecular nitrogen or as an oxide of nitrogen usually carried out by denitrifying bacteria.

Denitrifying bacteria

A group of soil bacteria which breaks down nitrates anaerobically to produce nitrogen. This happens under anaerobic conditions.

Dense fog advisory

Advisory issued when fog reduces visibility to a furlong or less. This weather forcast helps in avoiding the possible hazardous conditions.

Density

The mass concentration expressed as the mass per unit volume. In the micrometeological studies the bulk density, particle density etc., are the most commonly used terms in explaining heat transfer in the soils.

Depletion phase (irrigation)

Portion of the total irrigation time between inflow shut-off and beginning of recession at upper end of the field.

Depletion

The reduction in the intensity of incoming solar radiation in a crop canopy due to passage through the same from top to the bottom, because of reflection, absorption and scattering by plant parts.

Depression

In mesoscale and macroscale meteorology, an area of low pressure is commonly referred to as depression. Under certain other situations, a trough is also called as a depression. Also see 'cyclone'.

Depth of cut

The maximum depth of the penetration of the tool measured with reference to the initial soil surface.

Derecho

A line of intense, widespread, and fast-moving thunderstorms that moves across a great distance. They are characterized by damaging straight-line winds over hundreds of kilometres.

Desalinization

The process in which the salts either from soil or water are removed.

Desert crust

A hard layer containing calcium carbonate, gypsum or other binding materials, which is exposed at the surface in desert regions.

Desert soil

It often contains high percentage of soluble salts, high pH, poor in organic matter, water is scarce with recurrent drought and in this area usually millets are grown.

Desert

A region of extremely arid climate and the soils consisting of more sand aggregates. The moisture content is extremely low both in the soil as well as in the air in almost all seasons except during late nights. Cold winters and meager precipitation (< 250 mm) are the basic characteristics in desert areas.

Desertification

The process of spreading of desert conditions into adjoining semi-arid areas, which may be caused both by climate and human factors. The influence may be either long term or short term depending upon the human activities like overgrazing, deforestation and exhausting the already depleted soils with over cropping, etc.

Desilting

Deepening of the farm ponds or banks or reservoirs by removing the excess silt or clay for maintaining the optimum capacity.

Desiccation

The process of drying to remove moisture from an object to a definite level.

De-suckering (tobacco)

Getting rid of the suckers (side-shoots) either by removing manually or by use of chemicals. These side-shoots emerge due to removal of the tip of the plants with an objective of diverting nutrients to the most economical part of the plant, i.e., leaves

Determinate growth

One of the characters of new plant types of cereals is, flowering of plant species is uniform within certain time limits. The plant first completes its vegetative phase and then enters into reproductive phase, thereby making a clear cut distinction between these two growth phases.

Development of a crop

This is a physiological phenomenon which encompasses the activities of the crop resulting from growth and differentiation.

Dew point hygrometer

An instrument used to measure the dew point temperature. The same one can also be used to indicate the 'frost' depending upon the recommendations and specifications of the manufacturer.

Dew point or dew point temperature

The temperature to which a given parcel of unsaturated air must be cooled to produce a state of saturation at constant pressure and water vapour content. At this temperature saturated vapour pressure is equal to actual vapour pressure.

Dew

This is one of the forms of condensation. When air close to the Earth surface cools at night below its dew point, it gets saturated and water is deposited in the form of small droplets. Here, a point to underline is that either the Earth surface or the object on which the dew forms should not be below freezing point.

Diabatic process

A thermodynamic change of a system involving transfer of heat across the boundaries of the same system. Also see 'Dry adiabatic lapse rate' and 'Dry adiabat'.

Diara lands

The lands which are situated in between the natural levees that get inundated for different length of periods and are periodically eroded and formed by meandering, braiding and coarse change of rivers.

Dibbler

A device used for dibbling.

Dibbling

A method of sowing in which crop seeds are placed in holes made with help of dibbler wherein specific spacing and number of plants are maintained per unit area.

Dicalcium phosphate (CaHPO$_4$)

A product containing not less than 4 per cent P$_2$O$_5$ in citrate-soluble form, which is considered available to plants.

Diffluence

This word is the opposite of confluence meaning a rate at which wind flow spreads apart and long an axis oriented normal to the flow in question.

Diffraction grating

A device used to separate a beam of light into its constituent wavelengths.

Diffraction

The bending of light wave while passing through a slit.

Diffuse radiation

Radiation reaching the Earth's surface after having been scattered from the direct solar beam by molecules or suspensoids in the atmosphere.

Diffuse reflection

Some surfaces possesess the property of reflecting or returning the rays of light when they fall upon them. In diffuse reflection, a rough surface reflects light at many angles.

Diffuse solar radiation

The downward part of the solar radiation, which is scattered by the atmospheric gases and dust particles and is also reflected and transmitted diffusively by clouds and which passes through a unit horizontal area.

Diffused light

The 0.4 to 0.7 microns part of the scattered radiation.

Diffusion coefficient in soils

This is the ratio of the amount of water transmitted under unit potential gradient to the amount of water stored to raise the moisture potential by a unit. In this context it should be noted that, the molecules or ions of a dissolved substance move freely through the solvent, the solution becoming uniform in concentration.

Diffusion coefficient

The molecules of all gases move freely and tend to distribute themselves equally within the limits of the vessel enclosing the gas. As a principle, the rate of transfer of mass between two planes is proportional to the concentration difference which exists between the planes and inversely proportional to the cross sectional area of the medium through which the diffusion process occurs. In putting this in an equation form, the constant of proportionality is called as diffusion coefficient of the gas.

Diffusion equation

The differential equation that defines the process of diffusion in fluids.

Diffusion of gas

All gases diffuse within the limits of any enclosing walls. All gas molecules move from regions of high concentration to regions of low concentration until the concentration of gas becomes uniform and this process is known as diffusion of gases.

Diffusion of water vapour

This is the free movement of water vapour molecules within the crop canopy due to difference in densities. In the atmosphere also the same phenomena can be noticed. As per the Graham's law, the rate of diffusion of gases through porous bodies is inversely proportional to the square root of their densities. In many micrometeorlogical methods, water vapour molecules are considered as gas and the physical principles are applied to arrive at a proper result.

Diffusion pressure deficient

This is the synonym for 'suction force' and also to 'turger deficit'. This is basically a potential in all practical uses. In the water logged soils where the moisture potential is almost zero, the unlimited osmotic absorption of water through cell walls requires the expenditure of energy to reject water by the cell and to maintain its osmotic potential. As said above, this potential is called the diffusion pressure deficient.

Diffusion pressure deficit (DPD)

It is the difference in diffusion pressure of a solution and pure solvent at the same atmospheric pressure.

Diffusion

In a plant canopy the molecules of water vapour, heat and other gaseous components exchange energy with plant parts particularly leaves because of gradients of concentration. This is a continuous process because of the presence of eddies and their apparent random motion. This particular molecular agitation is known as the process of diffusion.

In meteorology, this term is used to denote the rate at which a cloud grows and individual elements in it become spaced further apart over a specified period. In environmental physics, this term is used as suffice to gases, liquids, etc., to indicate their movement.

Diffusivity

See 'Thermometric conductivity'.

Digestion

The conversion of complex foods, that are insoluble to simpler substances which are soluble in water.

Dike (levee)

An embankment to confine or control water, especially one built along the banks of a canal or a river to prevent overflow of low lands.

Dimensionless number

In certain micrometeorological field problems, some factors are often tried to determine whether the physical principle can be applied to them in that situation or not. To be on the safer side, under such situations, three to four observable and dimensionless factors are assembled and a fundamental parameter is finally found, to give more clarity to the situation. Such a parameter, naturally explains only some probable dynamic behaviour. Some examples are Reynold's number, Nassault number, etc.

Dimensions of units

For physical quantities, these are the powers to which the fundamental units, length, mass, time, temperature and charge are raised. For example the dimensions of velocity are length divided by time. Like this, different quantities have different units.

Direct drilling

The sowing of seeds directly in the field with a seed drill without soil preparation and which follows the weed control through other methods.

Direct solar radiation

The quantity of solar radiation received from the solid angle subtended by the Sun's disc and received on a unit surface held normal to the Sun's rays.

Directed spray

An application made to minimize the amount of herbicide applied to the crop. This is usually accomplished by setting nozzles low with spray patterns intersecting at the base of the plants just above the soil surface.

Directional Shear

The shear created by a rapid change in wind direction with height.

Discharge

The release of any quantity like electricity , water or conversion of one form of energy into another form, etc. But in all the cases, it should be expressed in proper units and at relevant rates.

Discontinuity

A zone of transition between two differing air masses in the atmosphere.

Disdrometer

An instrument used to measure the size of the diameter of the raindrop.

Disease endurance/tolerance

Capacity of plants to tolerate the attack of a disease causing organisms without exhibiting much damage and symptoms.

Disease resistance

Capacity of plants to withstand, oppose or overcome the attack of disease causing organisms. It is variable in amount ranging from zero to cent per cent and therefore, different terms as given below are used to denote the different degree of resistance. Absolutely free from disease (immunity), practical resistance: high or moderate resistance, susceptibility: very low or without resistance.

Disinfectant

A substance which can kill or inactivate pathogenic microorganisms in the near vicinity or on the surface of plant parts before infection.

Dispersed soil

Soil in which the clay readily forms a colloidal soil, usually with a low permeability, which upon drying, shrinks, cracks and becomes hard and upon wetting swells readily and is plastic.

Dispersion

The splitting of any light beam into its component wavelengths. For example, a beam of ordinary sunlight on passing through an optical prism is divided into light of different wavelengths of which it is composed.

Dissipation of energy

A more commonly used term in thermodynamics. In the relationship between the work done and the heat produced, on some occasions, the heat transforms into low grade heat, which may not be truly available form for any practical purpose. This transformation of the energy is known as dissipation of energy.

Disturbance

In agrometeorological terms, an area where wind, temperature, pressure, etc., show signs of cyclonic development.

Diurnal coefficient of viscosity

Diurnal means (a)daily (b)performed or completed once every twenty four hours. For two parallel layers in the direction of flow, the viscous force is proportional to the velocity gradient between the layers. The constant of proportionality is called the coefficient of viscosity of the fluid. So, with this information the diurnal coefficient of viscosity can be arrived at by dividing the tangential shearing stress with lateral shear, arising from viscosity. The S.I. units are Newton per second per metre per metre (for coefficient of viscosity).

Diurnal range

In micrometeorological studies of crop plants, the effect of any periodic quantity related to weather elements are studied, for 24 hours, which reoccur every 24 hours during the crop growth period. These studies help to integrate for any period of time according to the convenience and calculations. The upper and lower limits of 24 hours variation in weather elements thus studied is called diurnal range.

Diurnal variation

A rhythmic change in the observed value of micrometeorological parameters in a crop canopy during the course of twenty four continuous hours from one sunrise to another sunrise through only one sunset.

Diurnal

Daily, especially pertaining to a cycle completed within a 24 hour period, and which reoccurs every 24 hours. Diurnal measurement of weather elements in crop fields is a major activity in agrometeorology.

Divergence

This is a measure of expansion in velocity field, usually the horizontal velocity field. If a constant volume of fluid has its horizontal dimensions increased, it experiences divergence and by conservation of mass, its vertical dimension must decrease. Actually, divergence refers to these stretching components of the air in the local horizontal.

Diversified cropping

A cropping programme in which no single crop contributes 50% or more towards the total crop production or monetary income (comparable equivalents) annually.

Diversified farm

A farm on which no single enterprize or source of income equals 50% or more of the total receipt and on such farm the farmer depends on several sources of income.

Diversion canal

A canal to divert water from one point to another. In irrigation practice, it extends from the point of diversion at the main canal to the beginning of the distribution system.

Diversity index

It measures the multiplicity of farm products which are planted by computing the reciprocal of sum of squares of the share of gross revenue received from each individual farm enterprise in a single year. Thus, it is in effect an average of the number of crops harvested weighted in a particular way by the value of production of each of the crops.

Doldrums

The doldrums are regions of calm and light winds in the vicinity of the equator roughly between 5 degrees S and 5 degrees N latitudes. This belt of calms lies practically between the two trade winds.

Domain

Let A and B be any two sets and let F denote a rule which associates to each member of A, a member of B. We say that 'F' is a function from A into B. Also A is said to be domain of this function.

Domains

The clay aggregates in soil, which are readily useful to prepare a seed bed.

Doppler effect

The change in the colour of light or pitch of sound due to the motion of the source, the receiver or the medium. This is also known as Doppler shift or Doppler's principle and is named after D.J.Doppler (1803-1853). A classical example is that when light is emitted by a moving object it appears more red when it is receding from the observer.

Double cropping

A crop intensification programme in which two crops are taken in a farming year.

Down draft

A small scale downward moving air in the lee of a mountain or any other obstruction.

Down wash

On some rare occasions the air moves downward when obstructed by narrow constructions. The main reason for this phenomenon is the existing negative pressure created by the obstruction, in the flow of air. This rare process is termed as down wash. Here 'narrow constructions' is a relative term, as far as its application in crop micrometeorology is concerned.

Downburst

A severe localized downdraft from a thunderstorm or shower. This outward burst of cool or colder air creates damaging winds near the surface.

Downpour

A situation in which a heavy rain is occurring, as if hanging.

Downslope effect

The warming of an air flow as it descends a hill or mountain slope.

Downward longwave radiation

The Sun emits radiation over the whole of the electromagnetic spectrum from Gamma rays to radio waves. The downward long-wave radiation is the infrared radiation emitted in the downward direction not only by the Sun but also the radiating constituencies of the atmosphere like water vapour molecules, carbon dioxide, ozone, etc.

Downward terrestrial radiation

The terrestrial radiation reaching downwards i.e., back to the Earth. The various components of the atmosphere, particularly those existing in the lower troposphere are the main causes for radiation. This is also known as counter radiation of the atmosphere, because of which the Earth does not become totally cool.

Downward total radiation

In this term, 'downward' means 'incoming' and 'total' means both solar and terrestrial. When it is 'solar', it should include 'diffused' and scattered parts also. So, this is the total radiation directed towards the Earth.

Drag coefficient

This is the dimensionless ratio of the drag force on a leaf immersed in a fluid flow to the adjacent fluid momentum flux. The drag coefficient expresses the actual force on the leaf as a fraction of the maximum possible force of the fluid. In reality, it is convenient to express the combined effect of drag and skin frictions of the maximum possible force of the fluid. In reality, it is convenient to express the combined effect of drag and skin frictions on a body and define a total drag coefficient per unit cross sectional area.

Drag

The resistance offered by a crop canopy to the motion of air through it. For this resistance frictional or viscous forces are responsible.

Drainage water

The excess water that the soil is unable to hold against the force of gravity. Also water from surface, ground or storm water flowing in to a drain.

Drainage wind

A katabatic wind which is caused by the cooling of air along the slopes of a mountain.

Drainage

The removal of excess water from surface or sub-surface land for good crop growth.

Drifting Snow

Snow particles blown from the ground by the wind to a height of less than six feet.

Drilling

The method of sowing seeds in rows at uniform rate and at controlled depth with or without covering them with soil.

Drizzle

This is one of the form of precipitation in which the size of the droplet is lesser than half a millimeter or 0.02 inches. Like rain, snow, sleet, etc., this is also a form of precipitation, but the intensity is very light and the fine droplets of water barely reach the   ground.

Dropsonde

A radiosonde dropped with a parachute from an aircraft rather than lifted by a ballon. It is released to measure the atmosphere upto troposphere from the surface.

Drought resistance

The inbuilt mechanism of plant to maintain growth and yield with little or no major set back under moisture stress conditions.

Drought

The condition under which crops fail to mature because of insufficient supply of water through rains. A situation in which the amount of water required for transpiration and evaporation by crop plants in a defined area exceeds the amount of available moisture in the soil.

Dry adiabat

line describing a dry adiabatic process on a thermodynamic diagram. This diagram may be a tephigraph or any other standard diagram.

Dry adiabatic lapse rate

The rate of decrease in temperature with increase in height of a parcel of dry air when lifted adiabatically. A situation is said to be adiabatic when there is no exchange of heat with its surroundings and the atmosphere is in hydrostatic equilibrium. Unlike environmental lapse rate the dry adiabatic lapse rate remains the same at the rate of a decrease of 10 degrees centigrade for every 1000 metres of increase in height.

This is, basically, an adiabatic change in a rising parcel of a dry air. A decrease of 0.01 degree Celsius per metre with increase in one metre height of a parcel of dry air when lifted adiabatically.

Dry adiabatic

In meteorology, the dry adiabatic lapse rate is said to be a process lapse rate which is the rate of decrease in temperature with height experienced by an air parcel being lifted adiabatically through atmosphere in hydrostatic equilibrium. The term 'Dry adiabatic' refers to the describing of an adiabatic movement or process of an air parcel in which there is neither evaporation nor condensation of the water 'substance'.

Dry air

Air of low relative humidity in which there is no water vapour.

Dry bulb temperature

The temperature observed in a dry bulb thermometer of a psychrometer. This is useful not only to indicate the ambient temperature but also to compare the same with the wet bulb thermometer reading of the same instrument.

Dry bulb

A name given to an ordinary thermometer used to determine temperature of the air.

Dry farming/Dry land farming

The practice of crop production entirely with rain water received during the crop season or on conserved soil moisture in low rainfall (< 750 mm) areas of arid and semi-arid climate. The crops may face mild to very severe moisture stress during their life cycle.

Dry matter content

A substance obtained without water after subjecting it for oven drying at 65 ^{o}C till weighs to content figure.

Dry season

A period during which water deficiency occurs, as stored water used for evapotranspiration which falls below potential evapotranspiration.

Dry slot

An area of occupied by dry air that wraps into the southern and eastern sections of a synoptic scale or mesoscale low pressure system.

Dry Spell

A prolonged rainless period of at least 15 days during which not a single rain received. A relative term associated with drought. If dry weather conditions are less rigorous than those of drought (absence of rainfall) then that period is a dry spell.

Dryer

A unit which provides the conditions for reducing moisture generally by forced ventilation with or without addition of heat.

Drying

A reducing of moisture in a product usually to some predetermined moisture content of 10 to 12% that is safe for storage purpose.

Dune

A mound or ridge of loose sand piled up by the wind, usually occurs in desert areas.

Dung

The semi-solid excrement of animals used as manure and soil conditioner as well as fuel. The burnt material is mixed with seed for storage purpose.

Duality

The behaviour of light is described both by wave theory and particle theory and volumes of information is available on the subject. Based on these theories, sophisticated equipment is also being manufactured and are found to give the real measurements. So, the expression of these phenomena, by these ways is known by the term 'Duality'.

Dust devil

The swirling dust observed on a hot day over the land. In Andhra Pradesh State, these are seen during April and May every year. In the vortices of these systems considerable quantities of dried straw and grass are lifted which causes a lot of financial loss to the farmer. So, these are considered as abnormal weather situations in this State. In Rajasthan, near deserts, these are commonly seen.

Dust mulch

A shallow layer of loose surface soil. This is a management practice which is usually performed under drought conditions to reduce evaporation losses by arresting capillary movement of water.

Dust storm

An unusual, frequently severe weather condition characterized by strong winds and dust filled air over an extensive area. Here, extensive area is limited to local area.

Dust wheel index

A scale which is used to rank the volcanic dust.

Dust whirl

A small, intense, vertical disturbance, usually a few yards in diameter, in which large volumes of dust and debris are carried upward. This is also known as dust devil and usually occurs in arid and semi-arid regions of the world.

Duty (water)

Is the area irrigated by one cusec discharge of water during the crop period. It is equal to twice the base divided by delta.

D-Value

The deviation of actual altitude along a constant pressure surface from the standard atmosphere altitude of that surface.

Dynamic model

A crop growth model could be used to extrapolate research results from the regions that have been studied to other similar areas. There are several models in use in agriculture. Of them, the dynamic models contain the time variable explicitly unlike static models.

Dynamics

A part of physics , which deals with the forces producing motion. In other words, dynamics is the study of the motion of the objects in terms of forces which cause the motion.

Dyne

The unit of force. This is equal to the force required to give one gram mass an acceleration of one centimetre per second per second. One dyne is equal to 10^{-5} newton.

E

Earth moving

Ploughing and transport operation utilized to loosen, load, carry and unload soil.

Earthing up

The process of scouping the soil just near to the base of stems of certain crops like sugarcane, groundnut, cassava, papaya, potato, banana etc. to give support to the plants as well as to limit the incidence of certain insect pests too.

Earthlight (Earthshine)

The faint weak illumination of the dark part of the moon's disk produced by sunlight reflected onto the moon from the earth's surface as well as atmosphere.

Earthquake

A sudden, transient motion or trembling of the crust of the earth resulting from the waves in the earth caused by faulting of the rocks, volcanic activity or movement of tectonic plates etc.

Easterlies

The broad patterns of persistent winds with an easterly component. Example Easterly trade winds.

Easterly wave

- A wave like disturbance, of the tropospheric easterlies of low latitude. These are synoptic scale and occur in the region of intertropical convergence zone.
- An inverted, migratory wave-like disturbance or trough in the tropical region that moves from east to west, creating only a shift in winds and rain. The low level convergence and associated convective weather occur on the eastern side of the wave axis. It moves slower than the atmospheric current in which it is embedded and is considered a weak trough of low pressure. It is often associated with possible tropical cyclone development. It is also known as a tropical wave.

Easterly

Air mass moving from the east.

Eccentricity

The distance between the two foci divided by the length of major axis. This is a measure of the extent to which an eclipse is elongated, and this value is used to express the eccentricity of a planet orbit round the Sun.

Echo

The energy return of a radar signal after it has hit the target.

Eclipse

The passage of a non-luminous body into the shadow of another. This term is usually referred to indicate the darkening of Sun or Moon. The solar eclipse occurs when the Moon's shadow fall upon the Earth whereas the lunar eclipse occurs when the Moon enters the Earth's shadow.

Ecliptic

The great circle cut in the celestial sphere by a plane containing the orbit of the Earth around the Sun. It is inclined at 23.5 degrees to the equator at present.

Ecoclimate

Climate within crop environment is called ecoclimate.

Ecoclimatic adaptation

The acclimatization of a crop to the physical and environmental conditions of a locality.

Ecofarming

It is the potential for introducing mutually reinforcing ecological approaches to food production. It aims at the maintenance of soil chemically, biologically and physically the way nature would do it left alone. Soil would then take proper care of plants growing on it. "Feed the soil, not the plant", is the watch word and slogan of ecological farming.

Ecological crop geography

A branch of crop ecology which deals with the broad spatial distribution of crop plants and the rationale of such distributions in terms of physical and socio-economic environment influencing the production of crops.

Ecological equivalence

The situation in which two or more species can replace one another in the same ecological niche because of the similarity in their ecological amplitude.

Ecological factor

A part or element of climate or environment that influences the crop growth. An ecological factor may be either biotic or abiotic.

Any part or condition of the environment that influences the life of one or more organisms existing in an area under study.

Ecological niche

The capacity of a crop variety to exist in the ecosystem.

Ecology

- The study of the mutual relations between organisms and their environment.
- This word is derived from Greek language. 'Oikos' means home or surrounding and 'logos' means study. Therefore, ecology is the study of organisms i,e,, plants and animals in relation to each other and to their non-living environment.

Economic viability

Economic viability of a pattern can be determined by a budget analysis. The analysis includes cost of labour and purchased inputs for all operations specified as well as yields obtained. The profitability and returns to resources (productivity) of the pattern can then be compared with those of the existing cropping system to be replaced.

Economic yield of a crop

The biological yield times the harvest index of a crop.

Ecosystem

The part of the Earth's atmosphere in which life can exist is ecosphere. The collection of organisms that interact or have the potential to interact alongwith the physical environment in which they live is called an ecosystem.

A natural unit which consists of biotic communities and their abiotic environment, eg. Grassland ecosystem. Now-a-days large scale watersheds are categorized as ecosystems, because, the cost benefit implications of various inputs are considered for increased crop production.

Edaphic factors

This term is often used by agrometeorologists to represent both biotic and abiotic factors of the soil. This term is in common usage in both theoretical and applied aspects of ecological studies.

Eddy diffusivity

The exchange coefficient for the diffusion of a conservative property by eddies in a turbulent flow of fluid. Also see 'Eddying.

Eddy flux

The rate of transport of flux of a fluid property by eddies.

Eddy shearing stress

A stress applied to a body in the plane of one of its faces is known as 'shear'. A force per unit area is 'stress'. When stress is applied to a body, within its elastic limit, a corresponding strain is produced, and the ratio of stress to strain is a characteristic constant of the body. The shearing stresses due to turbulent eddy transfer to momentum is called as eddy shearing stress.

Eddy spectrum

The distribution of kinetic energy among eddies of various frequencies or sizes in a turbulent flow.

Eddy transfer coefficient

Same as coefficient of eddy diffusivity. A factor or multiplier which measures some specified property of a given substance, and is constant for that substance under given conditions is known as coefficient. The transfer of momentum because of the disturbance is known as eddy transfer.

Eddy velocity

This term has been used with various meanings at different situations in micrometeorological studies of crop plants. One version is that, in the aerodynamics, this is the difference between the instantaneous wind velocity at a point and the mean wind velocity taken at a point and the mean wind velocity taken over a given time interval. In other situations, while measuring the velocity profiles in crop canopies, this term is being used by the micrometeorologists to denote the velocity of the wind because of eddies, caused by both temperature and pressure gradients.

Eddy viscosity

The property of a fluid whereby it tends to resist the relative motion within itself is known as viscosity. The molecular momentum transport in a laminar flow is known as eddy viscosity. So, eddy viscosity is the apparent viscosity arising from turbulence.

Eddy

Any disturbance, deviation or fluctuation of fluid from the mean state of a turbulent fluid flow. A translating and rotating air parcel is often called an eddy. The transient eddies move in space (e.g., cyclones) while stationary eddies remain in fairly fixed locations (e.g., anticyclones).

Eddying

This is a product of turbulence. A mass of air like a small whirl pool retains its identity and size for varying short periods and moves forward horizontally with the velocity of the wind. This occurs only because of eddying.

Edison effect

The effect caused by emission of electrons from a hot metal. Named after T.A.Edison (184-1931).

Effective field capacity (tillage)

The actual area covered by the implement based on its total time consumed and its width.

Effective irrigation

Is controlled and uniform application of water to crop land in required amount at required time, with minimum cost to produce optimum yields without waste of water and adverse effect on soil in form of soil salinity and water-logging problems.

Effective outgoing radiation

The difference between the outgoing long wave terrestrial radiation and the incoming long wave radiation from the atmosphere. Here. The term 'terrestrial radiation' means the radiation of the Earth's surface only. The effective outgoing radiation is also known as effective terrestrial radiation. The units are watts per metre per metre.

Effective precipitation

The portion of precipitation that remains on the foliage or in the soil that is available for evapotranspiration and reduces the withdrawal of soil water. It is to be noted that this differs from effective rainfall in a way that, there are different forms of precipitation.

Effective rainfall

The rain useful for meeting crop water requirements excluding deep percolation, surface run-off and interception losses. The water thus received from rain shall be used during the crop growth period and must be available for all plant physiological process of the crop. The units are millimeters per period. Here 'period' means either 24 hours or days or 30 days, etc.

Effective rooting depth

Soil depth from which the full grown crop extracts most of the water needed for evapo-transpiration.

Effective soil depth

Depth of soil upto which plant roots penetrate to draw water and plant nutrients.

Effective temperature

The temperature above a certain minimum at which physiological processes such as growth of a crop plant, etc., are considered active. For many crop plants 5 $^{\circ}$C is considered to be the effective temperature. Even though, this is used as synonym to threshold temperature, both differ in their degree of applicability.

Effective terrestrial radiation

The difference between the outgoing long wave terrestrial radiation and the incoming long wave radiation from the atmosphere.

Efficiency

The ratio of the useful energy to the total input energy expended. This is expressed as percentage and the efficiency never be greater than unity.

Efficient cropping zone

It is a cropping zone in which relative yield index is 200% and relative spread indices vary from 90-200%.

Ekman layer

Under certain circumstances, in the planetary boundary layer, the vertical transfer coefficient is constant. This exists in the topmost layer of the atmospheric boundary layer, between the free atmosphere and the surface boundary layer.

Electromagnetic radiation

The radiation consisting of waves of energy associated with electric and magnetic fields, resulting from the acceleration of an electric charge. These waves travel at the speed of light and are capable of passing through empty space. These traveling waves of electric and magnetic disturbances are at right angles to each other and to the direction of propagation. This electromagnetic radiation is emitted by matter in discontinuous units called photons and the nature of this radiation depends upon the frequency.

Electromagnetic waves

The radiation consisting of waves of energy associated with electric and magnetic fields, resulting from the acceleration of an electric charge is known as electromagnetic radiation. These electric and magnetic fields propagate through space. A system, in which these fields vary, regenerating each other and traveling through space on a wave is known as electromagnetic wave. Basically, these are transverse waves which are caused by oscillations of electrons or groups of electrons.

Electromotive force

The source of electrical energy required to produce an electric current in a circuit. This is the rate at which electrical energy is drawn from the source and dissipated in a circuit when unit current is flowing in the circuit.

El-Nino

This is a Spanish term. In Peru, the climate is generally dry and cool, even though it is considered as a tropical country. The striking characteristics of this weather is unusually warm and associated with frequent rainfall. This situation occurs once in several years in the country. This peculiar phenomenon is also being taken into consideration in forecasting the behaviour of monsoon in the Indian sub-continent.

Elsasser's radiation chart
So many radiation charts are in effective use to solve radiative transfer problems. The one that was developed by W.M.Elsasser is based on graphs and the solutions obtained by using this method are graphical solutions only.

Emergence
Germinating seedling breaking through the soil surface.

Emerging farming
A concept involving the farming of fast growing plants/trees for the purpose of providing biomass that can be used directly as fuel or converted into other forms of fuel or energy products.

Emissive power
The total energy emitted from a unit area of a surface of a body per second. The total emissive power depends upon the temperature of the body and the nature of its surface. For instance, the emissive power of soil is the power of the soil to emit energy in the form of heat, either by conduction, convection or even by radiation process.

Emissivity
This is the ratio of the emittance of a given surface at a specified wavelength and emitting temperature to the emittance of an ideal black body at the same wavelength and temperature. Here, the word 'black body' is indicated only to impart specific meaning, such that, the same is the body with maximum possible emitting intensity at a given wavelength and temperature. So, it can safely be stated, finally, that the emissivity of a surface other than a perfectly black body is always less than one.

Emittance
This is the radiant flux density emitted by a unit area of a radiating surface.

Empirical model
In characterizing a given region regarding the possibility of growing a particular crop growth model is useful. In this direction, the empirical models are developed based on observed qualitative relationships which set out principally to describe.

Endemic (a plant or animal)

Confined to a particular area either because it is evolved there and has not been able to spread more widely or because it had survived there and no where else.

Energy and mass conservation

The law of conservation of energy states that in any system energy cannot be created or destroyed and the law of conservation of mass states that in any system matter cannot be created or destroyed. The general principle of the conversation of mass and energy states that, in any system the sum of the mass and energy remains constant.

Energy efficiency index

It measures the rate at which the use of energy inputs generates energy outputs. It is comparable to the economic efficiency index.

Energy flow

The intake, conversion, and passage of energy through a crop canopy.

Energy of gas

In the kinetic theory of gases it is assumed that there is no force of attraction or repulsion between the molecules of a gas. So, no work is done in changing the distance between the molecules and the gas possesses no 'potential energy'. The internal energy of a gas is only 'kinetic energy'.

Energy productivity ratio

Ralations showing food yield per unit of energy input and are most often used in evaluating the energetics of agricultural systems. The energy productivity ratio is food production in nutritional equivalents to fossil-fuel-based feed back.

Energy

The capacity to do work. This is equivalent to the work done by one dyne of force acting through one centimetre of distance. Meteorologically, the important types of energy are Kinetic, gravitational, potential, radiant heat, etc.

Enthalpy

The sensible heat is often taken as enthalpy in meteorology. In thermodynamics, this is the internal energy plus the product of pressure and volume. When it is said 'sensible heat' it means the heat content of the substance, in this context.

Entisols

Soils that have no diagnostic pedogenic horizons. They may be found in virtually any climate on very recent geomorphic surfaces.

Entrainment drag

The effective drag on a rising thermal. Also see 'Entrainment'.

Entrainment

Incorporation of surrounding air by a rising thermal, as with a rising cumulus cloud or an inversion layer.

Entrophy

This is the thermal property of a body which remains constant during an adiabatic process when no heat energy is given to or removed from the same body. So, this is a measure of the randomness or disorder of the object of a system. This is more or less internal and unavailable energy of a system, resulting from internal movement of molecules.

Environment

The sum total of all the external forces or factors both biotic and abiotic that affect the physiological behaviour or performance of an organism or of a group of organisms.

Environmental clock

The influence of environmental factors to initiate processes or activities of plant. For example, the influence of length of the day in flowering of certain crops.

Environmental resistance

The prerogative of the environment in which the restriction is caused by environmental factors upon the increase in numbers of individuals in a population.

Environmentally sensitive areas

Areas of high risk and wildlife interest designated under Indian environmental control regulations.

Epoch

This is a measure of geological or climatic time upon the long term scale,

Eppley phyrheliometer

An instrument used to measure the solar radiation.

Equation of motion

The equation of relative acceleration with net force acting on an air parcel of unit mass. This is a form of second law of Newton which states that, the rate of change of momentum is proportional to the applied force and takes place in the direction in which the force acts.

Equation of state

The equation relating pressure, volume and temperature of a parcel of an ideal gas or mixture of ideal gases.

Equation of time

The difference between the local apparent time and local mean time taken in the algebraic sense of 'apparent'. The time of rotation of the Earth upon its axis is not exactly equal to the time from noon to noon, the difference being caused by the motion of the Earth.

Equatorial air

An air mass which is warm because of its existence over the equator for several days.

Equatorial calms

Between the trade winds of the southern and northern hemispheres sometimes a zone of calm wind exists. However, on some rare occasions slight variations may also be noticed, but certainly the situation is by and large calm.

Equatorial climates

The summation of weather conditions experienced near the equator which are characterized by high temperature, high humidity, heavy rainfall with little or no seasonal variations.

Equatorial trough

Usually, equatorial waves are also called as easterly waves. The reason is that, the equatorial air masses are transported into the equatorial zone by the trade winds and become stagnant for considerable durations depending upon the characteristics of the surface. Sometimes, few low pressure belts are tucked in between tropical high pressure belts.

Equatorial

The great circle of the Earth, lying in plane perpendicular to the axis of the Earth which is equidistant from the two poles is known as equator. The air originating near this equator is called as equatorial.

Equilibrium

A state of balance between opposing forces or effects. In a situation, thermal equilibrium exists when all parts of the system are at the same temperatures. So, in equilibrium condition, all parts of a sytem remain in their existing state.

Equinox

The astronomical moment at which the daylight and night are equally long, which occurs when the Sun is directly overhead.

Equivalent potential temperature

The temperature which a body possesses by virtue of its position is termed as potential temperature. The temperature of an air parcel, when all its water vapour is condensed by adiabatic decompression is known as equivalent potential temperature.

Eradication (disease)

A method of disease control in which the pathogen is eliminated after it has established.

Erodibility

The lack of resistance of a soil to soil erosion.

Erosion resistant crop

A crop, because of its dense foliage, root system etc., provides effective protection against soil loss by erosion.

Erosion splash

A form of soil erosion resulting from soil splash under the impact of falling rain drops.

Erosion

The detachment and movement of particles of the land surface by water, wind or other forces. Depending upon the magnitude of the problem it is devided into rill, sheet and gully erosion.

Erosivity

The ability of rainfall to cause soil erosion.

Essential elements

Elements obtained by a plant from the soil, air and water without which plant can not complete its life cycle or the elements can not be substituted by others that have direct or indirect function in plant metabolism.

Establishment

The successful growth of a crop in a new environment and place. After the sowing of a crop in a field, if the emergence is optimum then the establishment of the crop is considered as 'good'.

Estivation

The condition in which a crop plant may pass the summer in which its normal activities are greatly curtailed or temporarily suspended or dormant.

Eutrophic

This term is used to describe a body of water (a lake, pond or tank) with an abundant supply of nutrients and high rate of formation of organic matter by photosynthesis. Pollution of a lake by sewage or fertilizers render it eutrophic, which stimulates the excessive growth of algae.

Evaporation opportunity

Evaporation is the process in which the conversion of liquid into vapour takes place, without necessarily reaching boiling point. The opportunity for this process to take place is known as evaporation opportunity, which is the ratio of the actual amount of water evaporated within the atmosphere to the evaporative power. It should be noted that, the air is the main elements of weather which controls the evaporative power at any given situation. Perhaps, the temperature is also important, but it is only next to the characteristics of the air.

Evaporation pan

An instrument used to measure the evaporation.

Evaporation rate

The volume of liquid or solid water evaporated per unit area in unit time.

Evaporation

The phase changing of a substance from liquid to gaseous or vapour state. This is a physical process and considered as the reverse of condensation for all practical purposes.

Evaporative power

An important point which a micrometeorologist should remember in the studies of evaporation is that, when the fastest moving molecules escape from the surface of a liquid during evaporation, the average kinetic energy of the remaining molecules is reduced and therefore the evaporation cause cooling. Here, the evaporative power is the same as the weather of a region is favourable to the process of evaporation from a land which is uncovered by vegetation. When it is said 'weather' here, it means basically 'air'.

Evaporograph

An instrument used to measure the evaporation over a given period of time.

Evapotranspiration

The total process of water transfer into the atmosphere both by 'evaporation' of liquid or solid water from the surface of the Earth and 'transpiration' from the plants in a crop canopy. This is a specific term confined only to a land surface which is covered either with vegetation or inside the crop canopy. It depends upon the climate, crop and soil factors.

Evapotranspirometer

An instrument used to measure the evaporotranspiration.

Exchange coefficient

The ratio of the flux of a quantity in a given direction to its gradient in that direction. This clearly states that the mean vertical flux of any conservative quantity is proportional to the gradient of the mean of the quantity in the vertical direction. The factor of proportionality between the flux and the gradient is also known as exchange coefficient. For example, in case of light this is a measure of the space of diminution of any transmitted part of light.

Exchangeable cation percentage (ECP)

This term indicates the degree of saturation of the soil/exchange capacity with a cation, and is expressed as follows: milliequivalents / 100 grams of soil.

Exchangeable phosphate

The phosphate anion reversibly attached on the surface of the solid phase (anion-exchange capacity) of soils, in such form that it may go into molecular solution by anionic equilibrium reactions with other anions of the solution phase.

Exchangeable potassium

The potassium which is held mainly by the colloidal portion of the soil and is easily exchanged with the cation of the neutral salt solutions less the water soluble potassium. It is readily available to growing plant roots.

Exchangeable sodium percentage (ESP)
Is the degree of saturation of the soil exchange complex with sodium expressed in percentage

Exothermic
If heat releases in a system under study, that system is called exothermic.

Exotic (plant)
A newly introduced plant not native to a place.

Experimental design
A logical structure of an experiment, that helps in obtaining results with some precision.

Experimental yield
The yield level of a crop variety obtained at experimental stations where yield maximisation is the major objective using all possible sophisticated technology.

Experiment
A systematic procedure for making observations under controlled conditions, in such a way that they can be used for arriving at general conclusions regarding the population under study.

Exponential growth
In mathematics, the number indicating the power of a quantity is known as exponent. In micrometeorology, the growth of a quantity for which the gradient is directly proportional to the quantity itself is known as exponential growth. In the similar way, the decay is also known as exponential decay.

Exposure of instruments
In any meteorological observatory, the instruments shall be placed in such a way that these are not exposed to direct solar heat, shadow, water logging condition, etc. To indicate the exact state of any particular weather element, any instrument should be exposed so that it is the representative of the real state of that element which it shows.

Extended forecast

The weather forecast valid upto a period of five days from the date of issue of the same. The main emphasis is given from prediction of change of weather type to weather hazards like strong winds, dry and wet spells, etc.

Extensive property

The property of a substance which cannot be used to characterize the same because it varies with the amount in the sample. For instance, the mass and volume of soil in a sample is a classical example for this property, because of the fact that, with a slight change in the volume, the property will automatically change for the soil.

Extinction coefficient

A measure of the amount of light absorbed by a substance in solution. This is a very valuable term in crop micro meteorological studies. In crop canopies this is a measure of the space of diminution of any transmitted light and is the ratio between the light loss through the leaf to the light at the top of the leaf.

Extinction

The attenuation of the intensity of a radiation beam, as it passes through the crop canopy, by the absorption and scattering of plant parts. Also see 'Extinction coefficient'.

Extratropical cyclones

A cyclone outside the tropics. As compared to tropical cyclone an extra-tropical cyclone is greater in scale. Sometimes, one or two fronts may also be present.

Areas of low pressure which develop in the westerly wind belts of middle latitudes. The associated weather is unsettled with much cloud, strong winds, periods of rain and sudden changes of temperature as the frontal zone pass. The surface of the extra-tropical cyclone consists of an area of low pressure surrounded by winds blowing in an anticlock-wise direction in the northern hemisphere and in a clock-wise direction in southern hemisphere.

Extreme temperatures

For any season or during a required period, the highest and the lowest temperatures recorded as period the IMD guidelines, in India.

Eye wall

An organized band of convection surrounding the eye, of a tropical cyclone. It contains cumulonimbus clouds, rainfall and very strong winds.

Eye

In meteorological terms, the roughly circular areas of calm or relatively light winds and comparatively fair weather at the center of a well developed tropical cyclone. A wall cloud makes the outer boundary of the eye. This could be seen in remote sensing photographs.

F

Factorial design

An experimental design in which at least two series of treatments are used being combined in every possible combination.

Fahrenheit temperature scale

A temperature scale having freezing point of water as 32 degrees and the boiling point equal to 212 degrees at standard atmospheric pressure. Conversion to Celsius scale is 1 degree Fahrenheit equals 1.8 degrees centigrade plus 32.

Fair weather

Fine sky. Appearance of cirrus or cumulus cover is in fair amount but not covering the sky totally.

Fall speed

Same as thermal velocity. See 'Thermal velocity'.

Fallow land

Land left unfarmed for one or more growing seasons to kill weeds, make the soil richer.

Fallow systems

It is the sequence of crop years with fallow years. Extensive fallow systems are shifting cultivation systems. Intensive fallow systems are bush fallow and grass fallow systems.

Fallow

It is the practice of allowing crop land to lie idle during a growing season to build up the soil moisture and fertility content so that a better crop can be produced the following year. It is usually worked periodically to control weeds and improve moisture infiltration.

Farad

The unit of capacitance and this is defined as the capacitance of a capacitor equivalent to one coulomb, at a potential difference of one volt between the plates.

Faraday's effect

Originally, this is restricted to light but later on to all the electromagnetic radiation. In meteorology, this is taken as follows. The plane of polarization of a radar pulse travelling through the ionosphere is rotated by the combined effects of the ionization and the magnetic fields of the Earth. So, the extent of the ionization in the ionosphere can be calculated by reflecting the radar pulses from the Moon or other Earth satellites and by measuring the total rotation.

Faraday's law of electromagnetic induction

Whenever the number of magnetic lines of induction or magnetic flux through a circuit changes, an induced electromagnetic force is produced in the circuit. The magnitude of induced e.m.f. is proportional to the rate of change of magnetic flux and lasts so long as the change continues.

Farm budgeting

A process of estimating costs, returns and net profit on a farm.

Farm enterprise

An individual crop or animal production function within a farming system which is the smallest unit for which resource use and cost return analysis is normally carried out.

Farm forestry

Farm forestry is a process in which trees are grown specifically for fuel, food and for a variety of other reasons on a farm. It is a "three dimensional farming" and provides an alternative to monoculture grain farming. The three dimensions are defined as the trees themselves, the harvest from the trees which is used to feed livestock and the animals and the animal products.

Farm irrigation efficiency

Is the ratio expressed in percentage of irrigation water available for crop production to that delivered at the farm head gate.

Farm management

The branch of agricultural economics which deals with the business principles and practices of farming with an object of obtaining the maximum possible return from the farm as a unit under a sound farming programme.

Farm planning

A process involving many decisions to be taken in respect of the kinds of crops to grow, rotations, mixtures, soil and water conservation practices to be followed and buildings, bullocks, machinery purchase etc.

Farm pond

A small body of water retained behind a small dam or held in a pit dug in the ground by water harvesting techniques.

Farm receipts

The total amount of money received from the sale of farm products, increase in inventories of livestock, feed supplies and receipts from outside labour, rent of building etc.

Farming system research

Is a highly location specific research with multi and inter disciplinary in nature and uses a whole farm approach for improved technologies to enhance and stabilise agricultural production. The research strategy includes base data analysis, on centre research and on farm research. This is the final evaluation of system in the real world situation of the farmer.

Farming systems

Farming systems represents, the cropping system and its interaction with other farmenter prizes viz., livestock, fisheries, forestry, poultry and the means available to the farmer to raise them for profitability. It interacts adequately with environment without dislocating the ecological and socio-economic balance on one hand and attempts to meet the national goals on the other.

Farmyard manure spreader

A machine to carry farmyard manure and spread it in a regulated quantity.

Fathom

The common unit of depth in the ocean. Which is equal to 6 feet or 1.83 meters. It can also be used in expressing horizontal distance, since 120 fathoms is equal to one cable or approximately one tenth of a nautical mile.

Feeder bands

The lines or bands of thunderstorms that spiral into and around the center of a tropical system. Which is also known as outer convective bands. A typical hurricanes may have three or more of these bands. They occur in advance of the main rain shield and are usually 40 to 80 miles apart. They are the lines or bands of low level clouds that move or feed into the updraft region of a thunderstorm, in its development.

Fencing

A device that fixes the boundary of the farm and helps to stop the encroachment of the land and protection from wild animals and cattle.

Ferrel cells

Wind cells in which air mass rises at the polar fronts in mid latitudes, which is the consequence of a thermally indirect circulation. This circulation is presumed to be driven by eddies. These eddies cause a weaker return flow towards the tropics and a stronger upper flow towards the poles. This can also be stated simply as the poleward air rises and equatorial air sinks.

Ferrugenous soil

Soil containing large amounts of iron minerals especially limonite and harmantite, characterized by red colour or shades of yellow and brown.

Fertiliser grade

An expression showing the legal guarantee of its available plant nutrients expressed as percentage of plant nutrients in a fertilizer, e.g. a 10-5-5 grade of fertiliser indicates 10 per cent nitrogen, 5 per cent phosphoric (P_2O_5) and 5 percent potash (K_2O).

Fertility (soil)

The ability of soil to supply all the essential nutrients in optimum amount and balance and in a form readily available to plants.

Fertilization (soil)

The application of fertilizers to the soil that aid in the nutrition of plants. A material in which declared nutrients are in the form of inorganic salts obtained by extraction and/or by physical and/or chemical industrial process (also termed mineral fertilizers).

Fertilizer broadcaster

A fertilizer distributor with a spreading width substantially greater than the width of the machine.

Fertilizer distributor

A machine which distributes fertilizer at regulated and selected rates.

Fertilizer drill

A machine to deposit fertilizer in soil at regulated and selected rates and at predetermined depth.

Fertilizer gun

A machine to throw a jet of fertilizer by mechanical, pneumatic or other means, usually fitted to the side of the machine.

Fertilizer mixture

A product obtained by physically mixing different fertilizers that are compatible so as to contain more than one of the three major nutrients, viz nitrogen, phosphorus and potassium.

Fertilizer requirement

The quantity of certain plant nutrient elements needed in addition to the amount supplied by the soil, to increase plant growth or crop yield to a designated optimum. The technical recommendation always expressed in Kg nutrient/ha, thus fertilizer required (kg/ha) =

$$\frac{\text{Nutrient recommended (Kg/ha)}}{\%\ \text{of nutrient in the fertilizer}} \times 100$$

Fetch

This is the distance of traversal by wind across a uniformly rough surface. While recording the micrometeorological observation, this can be taken for granted as the length of fetch area measured in the direction of the wind from the site in question. Here 'fetch area' means the area from which the wind flow is not obstructed and other micrometeorological parameters are also not so easily disturbed by surrounding ground situation. 'Site in question' means the area in which the actual measurements are taken. In meteorology also, this is the distance upwind, from the point of observation to a significant location. Here 'significant location' means a place where the standard synoptic anemometers are located in such a way that they can be seen with un-obstruction fetch or advantage. The word 'fetch' in conversation means the advantage.

Fertiliser ratio

The fertiliser ratio designates the relative proportion of three major plant nutrients, keeping the percentage of nitrogen as one in the ratio. Thus, a 5-10-5 fertiliser mixture which contains 5 per cent nitrogen, 10 per cent P_2O_5 and 5 per cent K_2O has a nutrient ratio 1:2:1.

Few

The amount of sky cover for a cloud layer between 1/8th and 2/8[ths], based on the summation layer amount for that layer.

Field capacity

Is the moisture content in percentage of a soil on oven dry basis when it has been completely saturated and downward movement of excess water has practically ceased, which usually takes place within 2-3 days after saturation. It is the upper limit of available soil moisture for plant growth. The moisture held at $-\dfrac{1}{3}$ atmospheres.

Field channel efficiency

Ratio of water made directly available to the crop and that received at the field inlet.

Field crops

Herbaceous plants grown on cultivated fields.

Field efficiency

It is the ratio of effective field capacity and theoretical field capacity expressed in percentage.

Field irrigation efficiency

Is the ratio expressed in the percentage of irrigation water available for use of crops to that delivered to a field.

Field irrigation requirement of a crop

The per cent moisture content in a field sample of soil at any one time.

Field moisture deficiency

The quantity of water which would be required to restore the soil moisture content to field moisture capacity.

Field strip cropping

A system of strip cropping in which crops are grown in parallel strips laid out across the general slope but which do not follow the contour. Strips of grass or close-growing crops are alternated with field crops, often.

Field systems

It is practised where one arable crop follows another and where established fields are clearly separated from each other.

Field water capacity

The maximum amount of capillary water that a particular soil is able to hold.

Fifteen atmosphere percentage

Is the moisture percentage on dry weight basis of a soil sample which has been wetted and brought to an equilibrium in a pressure membrane apparatus at-15 atm pressure (221 1b/sq inch). This characteristic moisture value for soils approximates the lower limit of water available for crop growth (permanent wilting percentage).

Fifty per cent yield decrement value

The measured value of the soil salinity or alkali that decreases crop yield by 50 per cent as compared with yields of the same crop on non-saline and nonalkali soils under similar growing conditions.

Filling

An increase of atmospheric pressure in the center of a 'depression' system. This is opposite to the term 'deepening' and more generally applied to a 'low'.

Final sample, laboratory sample

A representative part of the reduced sample or, where no intermediate reduction is required, of the aggregate sample.

Final water balance

Sum of all gains and losses of water over a given period of time; mm/period.

Fine texture

Consisting of or containing large quantities of the fine soil fractions, particularly silt and clay. Includes sandy clay, silty clay, and clay textural classes.

Fine weather

A blue and cloudless sky. Even if a few cirrus clouds are noticed, they donot express any vertical development.

Fire curing

A process of curing tobacco specially for tobacco used for chewing purpose. In this process leaves are wilted for a few hours in the field tied in to bundles and hung on laths in a smoke hut then smoked for 12 hours by burning.

Firewhirl

A tornado like rotating column of fire and smoke created by intense heat either from a forest fire or volcanic eruption.

First gust

Another name for the initial wind surge observed at the surface as the result of downdrafts forming the leading edge.

First law of thermodynamics

The law of conservation of energy is the first law of thermodynamics which states that in a system of constant mass, energy can neither be created nor destroyed. Mechanical equivalent of heat is the special case of this law.

Fitness of environment

The suitability of an environment or habitat for a crop to maintain its life, or for certain activities. The fitness of environment is an indication of the profitability of the crop.

Fixation of nitrogen

The fixation of atmospheric nitrogen in the soil by symbiotic and non-symbiotic bacteria.

Fixation of phosphorus

The conversion of soluble phosphorus nutrient in the soil into less soluble and unavailable forms.

Flag stage

The early post-emergence stage of onion seedlings between the "crook" stage and the emergence of the first true leaf. The bent tip of the seed leaf resembles a flag attached to a staff. Also referred to as the "knee" stage. In cereals refers to the stage when the upper-most leaf is out but the inflorescence is still in the boot.

Flaking line

A line of attached cumulus or towering cumulus clouds of descending height, appearing as stairsteps of the most active part of supercell.

Flash flood

A flood that rises and falls quite rapidly with little or no advance warning. A sudden and excessive rainfall, the thaw of an ice jam, etc cause these flash floods.

Floating plants

Type of plants that grows in water and remains in floating conditions, such as water hyacinth and water lettuce etc.

Floating rice

Tall rice grow where maximum water depth ranges between 1-6 m for more than half of the growth duration.

Flocculate

To aggregate individual particles into small groups or granules.

Flood irrigation

It is a method of irrigation where in entire land surface is flooded.

Flood plain

A strip of flat land bordering a stream or river, consisting of sediment laid down over the centuries by the river when it moves laterally within the valley walls and is subject to overflow during high floods . The Krishna, Godavari plains are classical examples for this.

Flood stage

The level of a river (or) stream where overflow onto surrounding areas can occur. This term is often used in weather forecast during floods to alert the affected people.

Flood

The overflowing of water from a river, stream, surface tank, etc., from their respective boundaries. On other specific instances, if water is accumulated because of poor drainage on crop fields, which are normally plain areas and inundate crops, then the situation is also said to be flooding.

Flop

In common usage, the word has a negative meaning where as in all branches of agricultural meteorology, this term is used to indicate a sudden change in climate. For instance, while recording micrometeorological observations, if a sudden cloud cover envelops in such a way that the recordings may not be true to type to those already recorded, then the situation is said to be flop.

Flora

1. The plants or micro organisms of a particular area
2. A descriptive list of plants in an area including key for identification.

Flotation

It is a process to remove light materials like bran from heavy materials like wheat by adding a flotation agent in the water.

Fluctuation

A relatively irregular departure from more normal or average conditions of the environment, in which a crop is grown.

Fluids

Generally, any substance with liquid characteristics is called a fluid. However, this term is used to denote any substance either a liquid or a gas which is able to flow, and take the shape of the vessel containing it.

Fluorescence

An interesting property to observe, in which a substance absorbs light of one wavelength and in its place emits light of another wavelength. The process ceases immediately after the source of light is cut off. It may be noted that in phosphorescence even after the source of light is cut off, the bodies emit light.

Flux density

This is the flux per unit area of a real or imaginary surface. In other words, for radiation falling on a leaf, the rate of flux of radiant energy through a square centimetre of leaf area, in a crop canopy.

Flux of energy

The normal flow of energy per unit time through a given surface is termed as 'flux'. This indicates that the flux is the rate of transport of a quantity like mass, energy, etc., across a given surface. So, the radiant energy, magnetic energy and other forms of energy including nuclear energy are being expressed in fluxes, for common understanding. In the energy balance studies of crop canopies, including crop weather models, this term is often used as radiant flux, sensible heat flux, latent heat flux, etc.

Flux of radiation

The term 'flux' for any vector quantity through an area is the product of the area and the component of the vector at right angles to the area. Flux of radiation is the radiant energy which crosses an area in unit time from all direction, which include the power emitted, transmitted or received in the form of radiation. To be precise, it can be stated as the radiant energy measured on the spot and as such in an area of a centimetre per centimetre per second. Here, the area may be a leaf, soil, space in between two plates in a crop canopy etc. When it is said 'the soil', it includes not only the surface layer, but in between the layers of soil into which roots of a crop penetrate and in a crop canopy at different convenient layers, flux can be measured.

Flyoff

This is a hypothetical and unique process by which the total amount of water escapes into the atmosphere in the form of vapour by the process of evapotranspiration from a crop canopy.

Fodder

Maize, sorghum or other coarse grasses harvested with seed and leaves and cured for animal feeding.

Fodder crops

The cultivated plant species that are utilized as livestock feed in the form of silage and hay.

Foehn

A strong warm and dry descending wind on the lee side of a mountain range produced by prior enforced ascent of air. It may be clearly noted that, this is a rare phenomenon, in terms of the warmness and dryness which is caused because of adiabatic compression upon descent. Generally, a mountain breeze is cool, but this is warm and dry which is seen in European Alps and a few other locations on the Earth.

Fog bank

A fairly well-defined mass of fog observed at a distance. Which is commonly seen at sea, over a lake, along coastal areas, etc.

Fog

This is a dense cloud in contact with the surface of the Earth. Some other scientists say that this is a hydrometeor consisting of numerous water droplets at the surface. This is formed when air is cooled to below its dew point and water vapour condenses around the constituents of aerosol. The visibility in fog is less than one kilometer. Also see 'Freezing fog'.

Fogbow

A whitish semicircular arc seen opposite the sun in fog.

Foliage

Also called leafage. The leaves collectively of a plant.

Foliar diagnosis

An estimation of mineral nutrient deficiencies, or excesses in plants based on examination of the chemical composition of selected plant parts, and the color and growth characteristics of the foliage of the plants.

Foliar fertilization

Fertilization of plants, or feeding nutrients to plants, by applying chemical fertilisers to the foliage usually in the form of spray.

Foliar spray

Any spray applied to plants when leaves are present.

Foot- candle

A unit of illuminance equivalent to one lumen per foot per foot. This is illuminance provided by a power source of one standard candle at a distance of one foot.

Foot-pound

A unit of work (energy). This is equal to the work done by a force of one pound weight acting through a distance of one foot.

Foot-pound-second system

A system of measurement in which foot is the unit for length and pound and second are the units for mass and time respectively.

Forage crop modelling and simulation

A 'model' is a simplified representation of a more complex reality in a forage-livestock production system. It describes how inputs or variables enter, move, interact and are controlled within the system and how the output of the system is generated and affected by other input and control variables.

Forage crop

Crops grown primarily for livestock feed, to be either harvested for hay, silage or green feed or harvested by grazing animals.

Forage forestry

The concept of forage forestry can simply be defined as "using the same land or a portion of it, simultaneously or sequentially for fodder, food and fuel.

Forage quality

Forage quality is defined as out put per animal and is a function of volunaty intake and digestibility of nutrients when forage is fed alone and to a specified animal. The terms like rate of consumption of digestible dry matter or digestible energy consumption have come into common usage these days to express forage quality.

Force

The work done by an external agency per unit of distance, on interaction with an object preferably by a pull. This is measured in newtons or pounds.

Forced convection

In the atmosphere, if mixing of relatively large volumes of air is due to mechanical forces such as deflection by large scale surface irregularities, turbulent flow caused by friction at boundary of a fluid, then the resultant heat transfer is said to be forced convection. The properties which influence the heat convecting ability are dynamic viscosity, heat capacity at constant pressure and thermal conductivity.

Forced turbulence

The essential feature of a turbulent flow is that the fluctuations are random. If much of the atmospheric turbulence is due to friction offered by the ground, then the turbulence is said to be forced turbulence. In these situations, the Reynold's numbers are occasionally high.

Forecast bulletin

The information in a statement of a clearly edited form on anticipated weather for a location, place, etc., for a definite period ahead, prepared as per the IMD guidelines.

Foreign water

The water from nearby area, to a place when there is no rainfall in the area and its immediate vicinity.

Forest

A plant community predominantly of trees and other woody vegetation usually with a closed canopy.

Forestry

The act, occupation or art of farming and cultivating forests and systematic utilization, reproduction and improvement of the productive capacity of trees in masses, including the planting and culture of new forests.

Form drag

The force experienced by the bodies in the direction of flow as a result of the decelaration of fluid, when they are immersed in moving fluids. It depends upon the shape and direction of the body. The maximum form drag is experienced by surfaces at right angles to the fluid flow.

Forms of radiation

There are two main forms of radiation, the short-wave radiation from the Sun and Long-wave radiation from both the Sun and the Earth and its surrounding atmosphere. Also see 'Short-wave, Infrared, Sky and Scattered radiations'.

Fortin's barometer

A mercurial barometer used to measure the pressure. Named after J. Fortin (1750-1831).

Foundation seed

The second link in the certified seed chain produced from breeder seed and handled in such a way as to ensure genetic identity and varietal purity.

Fractional interception efficiency

The fractional interception efficiency of a crop canopy is the derivative which is obtained by dividing the instantaneous intercepted PAR with instantaneous PAR flux density.

Fractional numbers

A set of numbers like p/q (e.g.., $\frac{2}{3}$, $\frac{3}{4}$) where p and q are natural numbers, was adjoined to the set N of natural numbers. This new set is known as the set and it obviously includes natural numbers as a sub-set and q being equal to 1 in this case.

Fractus

The elements of cumulus and stratus clouds that appear in irregular fragments. They look like shred or torn. A synonym of scud.

Free atmosphere

The gaseous layer surrounding the Earth is atmosphere, which is colourless, odourless, tasteless, etc. In meteorology, any element of weather which occurs as such is called a 'free element'. So a free atmosphere is the atmosphere in which there are no gradients as far as elements are concerned. The situation can be seen only above the atmospheric boundary layer and as such the term 'free atmosphere' should be applied to the atmosphere above the atmospheric boundary layer.

Free balloon

A balloon ascending freely in the atmosphere by usual buoyant forces.

Free convection

If mixing of comparatively large volumes of air is caused by density differences within the air which also results in temperature differences then the transfer is called free convection. This also occurs if motion is caused by density difference, with the fluids.

Free flow

Is a condition under which the rate of discharge is solely dependent on the length of crest and depth of water in the converging section of the parshall flume. In case of weirs the criterion is the depth of water at 4H (crest height) distance away from the weirs. There is no back pressure of water in a free flow condition.

Free ground water

Ground water in the interconnected interstices in the zone of saturation. It extends down to the impervious barrier, and moves under the influence of gravity in the direction of the slope of the water table.

Free lift

The actual lifting force of an inflated balloon and is expressed in grams.

Freezing drizzle

A drizzle, which is frozen on impact with the colder ground. It is reported as "FZDZ" in an observation and on the METAR.

Freezing fog

Fogs form because of the cooling of moist air below its saturation point. If super cooled water droplets freeze on impact with any solid surface then the fog is called freezing fog.

Freezing level

The level in the atmosphere at which the temperature is zero degrees centigrade and where the pure water is supposed to freeze.

Freezing nucleus

The components of an aerosol population which comprise particles on which the freezing of water occur.

Freezing point

The temperature at which a liquid changes to a solid under given conditions.

Freezing precipitation

Precipitation (rain) freezing upon impact with the ground

Freezing rain

Rain that falls as liquid and freezes upon impact to form a coating of glaze on the colder ground.

Frequency of waves

The number of waves (vibrations of wave motion) in unit time, usually one second, passing through a given point. The S.I. unit of frequency is hertz.

Frequency

For any cyclic motion the number of events per unit time. This is expressed in inverse seconds.

Fresh breeze

Wind speed between seventeen and twenty one nautical miles per hour or five on Beufort scale wind force.

Friability of soil

It characterizes the case of crumbling of soils. The moisture range in which soils are friable is also the range in which conditions are optimum for tillage.

Friction layer

The thin layer of atmosphere adjacent to the surface of the earth which is effective in slowing down wind upto approximately 1,500 to 3,000 feet above the ground.

Friction velocity

A basic and convenient wind parameter. The forces offering resistance to relative motion between surfaces in contact is called friction. So, friction velocity is defined as the square root of the turbulent horizontal wind stress in the surface boundary layer, divided by the air density in the surface layer. For all the wind profile studies in crop canopies, certain coefficients are being developed in crop weather models based on this parameter, at different stages.

Friction

See 'Friction velocity'.

Frictional turbulence

Same as 'Forced turbulence'.

Front

An interface between two different air masses of different properties like sharp gradients of temperature, wind, etc. A 'front' is a mesoscale phenomenon in the lower troposphere and is considered as a region of most rapid weather changes. The characteristics of air mass differ because of the fact that their origins are different, and a front is formed when these air masses meet. A front is called a warm or cold one depending upon the advancement of air masses. Also see 'Col-front, Warm front and Occluded front'.

Frontal lifting

The formation and simultaneous intensification of a front is called as frontogenesis. The temperature inversion is a common feature in this process and it is called as frontal inversion. The forced ascent of warmer air at a front brought about by convergence between two characteristically different air masses is called as frontal lifting. However, the dissipation of a front is frontolysis.

Frontal zone

A zone in which cyclogenesis is a routine process. This term is also applied to a zone of highest horizontal temperature gradient associated with a front.

Frontogenensis

- The initial formation of a front or frontal zone.
- The birth of a front, which occurs when two adjacent air masses exhibiting different densities and temperatures are brought by prevailing winds. It occurs most often along the eastern coasts.

Frontolysis

The destruction or disappearance of a front where the transition zone is losing its contrasting properties.

Frost resistance

The capability of a crop variety to survive the formation of ice crystals in their tissues. Irrespective of the stage of the growth of the crop, frost cause substantial damage. However, in majority cases young seedlings are prone to frost.

Frost

The deposition of ice crystals on a cooler land surface or objects, by diffusion and sublimation. This occurs in the atmosphere when air temperature and dew point temperature are below freezing level. A situation in which fog is thick enough to produce marked whitening of green grass is said to be near frost.

Frozen precipitation

Precipitation that reaches the ground in a frozen state. Examples snow, hail, etc.

Fujitapearson scale

This scale is developed by T. Theodore Fujita and Allen Pearson. Which classifies the severity of wind damage intensity based on the degree of destruction as it relates to the wind speed. It also takes path length and path width of the event into consideration. It is normally used to identify the most intense damage exhibited by a tornado.

Fully turbulent

Turbulent flow is the motion of fluids in which local velocities and pressures fluctuate irregularly. The commonly seen river and wind flows are said to be turbulent, technically. A fully turbulent situation occurs when the transport of atmospheric properties takes place by the bulk mixing of air due to molecular heat motion.

Fumigant

Chemical used in the form of a volatile liquid or a gas to kill insects, nematodes fungi, bacteria, seeds, roots, rhizomes or entire plants. Usually applied in an enclosure of some kind of in the soil.

Fumigation

This is used in different contexts. In agricultural entomology, the application of gas, smoke or smoke generating chemicals to control the unwanted insects, rats, etc., is called fumigation. In meteorology, the process of pollution, smoke and dust materials reaching the surface from the lower troposphere is known by this term.

Function

A quantity is said to be a function of another quantity when a change in one quantity results in a change in another quantity. It can be stated also as, for a change in one variable there is a unique change in the value of another.

In calculus a function is a rule which establishes a correspondence between two sets.

Funnel cloud

A tornado cloud or vortex cloud extending downward from the parent cloud but not reaching the ground. Technically, this is not a type of cloud.

Furrow crown

The peak of the turned furrow slice.

Furrow face

The vertical side of the furrow made in the soil, away from which the turned soil is thrown.

Furrow irrigation

A method of applying irrigation water to fields or orchards by small ditches or furrows which lead from the supply ditch.

Furrow planting

Planting in the bottom of furrows.

Furrow slice

The mass of soil cut, lifted and thrown to one side of an implement.

Furrow sole

The bottom of the furrow on which the plough bottom slides.

Furrow wall

It is an undisturbed soil surface by the side of a furrow.

Furrow

The trench formed by a tool in the soil during operation.

Fusion

In different fields of science, this term has different meanings. In micrometeorology, this is used to indicate the melting of ice or snow to form into water in crop canopies and making it available for plant physiological processes.

G

Gale warning

warning for marine interests for impending winds from 28 to 47 knots.

Gale

The wind with a speed between 32 and 63 miles per hour. On Beufort scale the numbers through 10. For convenience, in certain specific instances it is mentioned as the wind speed for which the 10 minutes average is 37 knots at height 10.

Galvanometer

An instrument used for detecting, comparing and measuring small electric currents.

Gamma rays

In the electromagnetic spectrum, the rays which have ionizing capabilities and those that can penetrate through the plant tissues. These rays are categorized under 'ultra-violet radiation' when full band is described.

Garden agriculture

Crops grown on land more or less adjacent to settlements. Usually vegetables etc., rather than staples, dependent on organic wastes from settlement for nutrient inputs.

Gas constant
The constant factor in the equation of state for perfect gases. This is equal to 8.3143 Joules per Kelvin per mole or 1.9858 calories per degree celsius per mole.

Gas
This is a state of matter which has no definite size or shape. It always occupies the whole of the space in which it is contained and the shape of the container is the shape of the gas.

Gaussian distribution
The distribution of variables characterized by a bell shaped symmetrical curve, for which the mean, mode and median are at the point of symmetry.

Geiger counter
An apparatus used to detect the radiations of radioactive elements. Named after Hans Geiger (1882-1947).

General circulation model
At present, there are different models in vogue, to explain the general circulation of the atmosphere. If the general circulation is modeled in a three dimensional procedure, then it is known as a general circulation model.

General circulation
The average general distribution of wind in motion on the Earth. This general circulation includes the scientific description of atmospheric dynamics not only statistically but also from the point of view of physics. The global wind systems like westerlies, trades, etc., are the examples of general circulation.

Genesis of soil structure
The causes and method of formation of the structural units or aggregates of soil.

Genotype
The heriditary make up of an individual plant organism which, with the environment controls the individual's characters.

Gentle breeze
Air in motion with a speed between nine to twelve nautical miles per hour or three on Beufort scale wind force.

Geomorphology
The study of the Earth's form and its evolution, both of which owe much to the action of water in rivers and glaciers. The NGRI is attempting to make this a full fledged science.

Geophysics
The study of the physics or nature of the earth and its environment.

Geopotential heights
The height above the sea level of a point, expressed in units of the potential energy of unit mass at that point. Since gravity varies slightly in the lower 99 percent of the atmosphere, geopotential and geometric heights are numerically interchangeable.

Geosphere
The portions of the earth, which includes the hydrosphere and the lithosphere.

Geostationary satellite
An orbiting satellite that maintains the same position over the equator during the earth's rotation. It helps in tracking the changes of weather.

Geotrophic wind
The wind moving parallel to straight isobars. These winds are characterised by a situation in which the coriolis acceleration is exactly maintained by the horizontal pressure gradient force. These winds exist approximately five to six hundred metres above the surface of the Earth. The wind for which the main characteristic is that it is balanced between the horizontal pressure and coriolis force.

Geotropism
The word 'tropism' means the curvature response of an organ to a stimulus. For example, the stem growing towards a source of light and roots away from the light. The word geotropism means a tropism in response to gravity. For instance, in groundnut crop peg penetration into the soil is called as positive geotropism.

Germination

Is the resumption of growth of a seed usually recognized by rupture of the seed coat or spore wall and appearance of the radicle and plumule from the seed.

Ginning (cotton)

The process of separating lint from seed cotton in ginning factory.

Ginning percentage (GP)

The weight of cotton lint obtained from seed cotton expressed in terms of percentage of seed cotton or percentage of lint obtained from 100 unit of seed cotton by weight.

Glaciation

The process in which a super cooled water cloud is converted into an ice cloud.

Glare

A harsh, uncomfortably bright light or reflection. During weather forecasting on television, glare should be avoided.

Glaze

An accretion of relatively clear ice. So, this is a coating formed on objects by freezing, when rain falls on objects or on the ground possessing sub-freezing temperature.

Global radiation

The sum of direct solar radiation from the Sun and the indirect diffuse sky radiation received on a unit horizontal surface.

Global solar radiation

The total solar radiation received on horizontal surface which includes the short-wave radiant energy emitted by the Sun's disc and the diffusively scattered solar radiation by the atmosphere components like dust particles, gases, clouds, etc. This is same as global radiation. Also see 'Global radiation'. The solar irradiance received on a horizontal surface which includes the direct component of sunlight plus the diffuse component of sky light . Unit is watts per metre per metre.

Gradient level

The degree of inclination of a slope, usually expressed as a unit rise in height per number of units covered along the slope. For small gradients the difference between the sine and tangent is small. In meteorology, the lowest level in the troposphere which is free of surface drag is known as gradient level. At this level the actual and gradient winds are not distinguished.

Gradient wind

The wind representing a balance of coriolis and centrifugal forces. Such wind results when the wind moves parallel to curved isobars, relative to the surface of the Earth.

Gradient

The rate of decrease of fluid property in a given direction. For example, in case of temperature gradient, a spatial rate of change in the direction of maximum rate of increase occurs.

Grain development

Is a continuous process where grain undergoes distinct changes before it fully matures i.e., Milk stage, soft dough stage, hard dough stage or full maturity stage.

Grain drying

Drying the harvested grain usually in sun to reduce the moisture content to an acceptable level. Drying process is basically the transfer of heat by converting the water in grain to a vapour and transfering it to the atmosphere. The purpose is to bring down the moisture contatn to a safest level to 12% to avoid the storage pests.

Grain grader

A machine used for grading the grain.

Grain legumes

Pulse crops belonging to family leguminoceae chiefly grown for grain purpose and consumed in various forms and mainly as 'dal'.

These crops belonging to the family leguminoceae which are grown for their edible seeds which are considered to be the cheapest source of protein compared to animal origin.

Grain processing

The process of upgrading grains for improving their marketability, storability and suitability for human consumption.

Grain separator

A machine to remove impurities from grain or other seeds and to sort them into two or more fractions.

Grain shattering

A term used to describe the loss of grain or seed from the ear head or spike while crop is still standing in the field at dead ripening stage and some how delayed for harvesting.

Grain

An indehiscent simple fruit of a grass family which grade along its length.

Gram equivalent

The mass of substance which will react with or is otherwise equivalent to one gram atom of hydrogen.

Gram

A unit of mass, equal to one thousandth of the International prototype kilogram. This is also equal to the mass of one cubic centimetre of water at 4 degree Celsius.

Grand growth

The period of enlargement of cell, tissue, organ, or organisms. This starts slowly immediately after differentiation, and increases to a maximum rapidly, finally falls of to zero.

Granular fertiliser

A fertiliser composed of particles of roughly the same composition and about one-tenth of an inch in diameter. This kind of fertiliser is superior in efficiency compared to the fine or powdery fertiliser, due to ease in handling and less fixation in soil.

Granular structure

Soil structure in which the individual grains are grouped into spherical aggregates with indistinct sides. Highly porous granules are commonly called crumbs. A well-granulated loamy soil has the best structure for most crop plants.

Graph of a function

The graph of a function defined by $y = f(x)$ is a pictorial representation of the function and can be obtained by locating on a rectangular coordinate system, the points defined by the number pairs (ax, y) or $[x, f(x)]$.

Grash of number

The ratio of buoyancy force times an inertial force to the square of a viscous force.

Grass minimum temperature

In a standard meteorological observatory, the minimum temperature recorded as per the IMD guidelines, on a thermometer, with its bulb touching the tops of leaves of short grass.

Grass

Botanically, any plant of the family Gramineae is called grass. The term 'grass' in the context of feeding livestock and grassland agriculture is not limited to the narrow botanical sense alone, but also includes their common associates of the legume family. However, the term does not include cereals when grown for grain.

Grassed waterway

A natural or artificially made way for the flow of water, usually shallow on which erosion resistant grasses are grown, to permit water to run-off fields thus reducing erosion where the crops are growing.

Grassland ecosystem

It pertains to the ecosystems of grass lands and pastures.

Grassland farming

Farming system that emphasizes the importance of grasses and legumes in livestock and land management in which the legumes are the keystones and grasses form the backbone of the grassland farming.

Grassland

The land on which graminaceous species represent the dominant if not the exclusive vegetation. Grassland is intermediate in status between forest or woodland on one hand and the desert on the other.

Graupel

A form of frozen precipitation consisting of snowflakes, ice crystals, supercooled water droplets, etc frozen together.

Gravitational acceleration

The downward acceleration of a body falling freely, as measured from a frame fixed to the surface of the Earth. This is also equal to the gravitational force per unit mass of the body. Gravity causes bodies to fall to Earth with a uniform acceleration, but the magnitude of the acceleration due to gravity varies with geographical location and altitude.

Gravitational field

Gravity is responsible for the weight of the body. The region in which one massive body exerts a force of attraction on another massive body is known as gravitational field. Here, the word 'massive' is relative.

Gravitational water

The free water which moves down into soil due to gravity until it reaches water table. As far as micrometeorological studies are concerned, the gravitational water is of no use to plants as it passes off before it is absorbed by roots of plants. On saturation of the cropped field, this component goes on increasing.

Gravity waves

It is a reality that the flow of fluids takes place due to several factors, of which temperature gradient, pressure gradient, rotation of Earth, etc., are the main causes in addition to the force of gravity. When a flow of fluid is disturbed predominantly due to gravity, then the disturbed waves are called as gravity waves.

Gravity

The force of attraction between the Earth and bodies on its surface. It is a fact that every particle in the universe attracts every other particle with a force which is directly proportional to the product of the masses of the particles and inversely proportional to the square of the distance between them. The term gravity also includes the force of Earth on bodies within its gravitational field.

Grazing capacity

Number of animals a given pasture will support at a given time or for a given period of time.

Grazing intensity

The number of animals per unit area of grazing land.

Grazing

The eating or partial defoliation of any kind of standing vegetation of domestic live stock or other animals.

Green flash

A brilliant green coloration of the upper edge of the sun.

Green house effect

The effect because of which, the Earth is warmed more than expected due to the presence of atmospheric gases like carbon dioxide and other tropospheric gases. It is a fact that the short-wave radiation can pass through the atmosphere easily, but the resultant outgoing terrestrial radiation cannot escape easily because the atmosphere is opaque to this radiation. This re-radiant energy acts to conserve the heat and this conservation is called as the green house effect, because of which the Earth is warm.

Green manure crop

A crop grown and ploughed down while still green normally flower bud initiation stage for improving soil properties.

Green manure in situ

A practice of ploughing or turning into the soil of undecomposed green manure crops in the same field where the crop is grown.

Green manure trampler
An impliment used to trample and press the green manure crop in the field.

Green manuring
A practice of ploughing or turning into the soil undecomposed green plant material for improving the physical, chemical and biological properties of the soil. In general legume crops are used for this purpose.

Greenleaf manuring
This refers to turning under of green leaves and tender green twigs collected from shrubs and trees grown on bunds, waste lands and nearby forest areas. The common shrubs and trees useful for this purpose are glyricidia (Glyricidia maculata), sesbania, karanj (Pongamia pinnata). Subabul (Leucaena leucocephala) Tephrosia sp, Anona sp etc.

Greenwich mean time
The local mean time at zero longitude. The zero longitude is the longitude of Greenwich.

Grey body
A body for which the monochromatic emissivity of the body is independent of wavelength.

Gross irrigation requirement (GIR)
The total amount of water applied through irrigation including losses during conveyance, distribution and application of water in the fields etc.

It is given by formula $\text{GIR} = \dfrac{\text{Net irrigation requirement}}{\text{irrigation efficiency}}$

Gross primary productivity
The total rate of photosynthesis including the organic matter used up during respiration in a given time by a crop stand.

Gross productivity
Total productivity of a system or plant per unit area and per unit time.

Gross return

Total income from the farm, by virtue of sales of entire farm produce.

Ground clutter

A pattern of radar echoes reflecting off fixed ground targets such as buildings or hills near the radar.

Ground cover

Protective vegetation such as provided by a solid-seeded stand of forage species or the residue from the previous crop.

Ground fog

Fog created when radiational cooling at the earth's surface lowers the temperature of the air near the ground to or below its initial dew point.

Ground heat flux

The rate of transfer of heat to and from the surface of the soil. In the crop radiation energy studies, this is considered as the amount of heat energy expended in heating the soil. This is measured with heat flux plates.

Ground water

Subsurface water in the zone of saturation that is free to move under the influence of gravity.

Growing degree day

*This is counted for each degree that the daily mean temperature is over and above the base temperature. Units are degrees Celsius.

*This is a concept and simple means of relating plant growth, development and maturity to air temperature. This is calculated by deducting the threshold temperature from mean daily temperature. These units are also called as heat units and this concept is useful in studying the suitability of a crop variety to a new environment.

Growing period

The duration, in days, of the period when both temperature and soil moisture permit crop growth (growing season, growth cycle).

Growth analysis of a crop plant

The study of integrated effect of environment on growth and development which is expressed mathematically.

Growth analysis

Method of analysing growth through plant sampling technique. By this method observations are recorded periodically on various plant characteristics (height, flowering time, ear bearing tillers etc.,) and also sampled at varying intervals (weekly or fortnightly) when the different parts such as stem, root, leaf, fruit, seed, etc., are separated, weighted and measured. From these the physiological parameters such as relative growth rate; crop growth rate etc., are calculated and these are further analysed in term of contributing component parameters like net assimulation rate and leaf area.

Growth correlation

The different parts of a plant do not grow in isolation but follow an orderly pattern due to control of one part of the plant over the other. This controlling effect exerted by one part over the other is known as correlative effect.

Growth curve

A graph in which the size of some plant characteristics (height, weight, leaf area etc.) is plotted against time or age. The rates of growth of any single cell or mass of cells are sigmoid (S-shaped), slow at first, increase with time to a maximum and then fall off at about same rate to zero. The period of maximum growth is called "grand period" of growth.

Growth factor

A specific substance that must be present in the growth medium to permit cell to multiply.

Growth of a crop plant

Growth of a crop plant denotes an irreversible increase in size by cell division and enlargement accompanied by synthesis of new material and organization of sub-cellular organelles.

Growth stages of cereal crops
1. *Tillering stage* : additional shoots are developing from the axillary buds
2. *Jointing stage* : when stem internodes are elongating
3. *Boot stage* : when leaf sheath swells due to the growth of developing spike or panicle
4. *Heading stage* : when ear head is emerging from the sheath and anthesis taking place
5. *Maturity stage* : when individual grain in the panicle ripens.

Growth

An irreversible process in which there is an increase in size, or dry weight or volume of an organism that happens due to cell devision, cell elongation and multiplication.

Grubbing

The process of removing stumps, sunken stones, boulders and all roots over a minimum size and upto a specified depth.

Guinea monsoon

The Indian sub-continent is bestowed with regular monsoon winds, through which the rainfall occurs and economy of the nation depends. In West Africa also, warm and humid southwesterly winds blow between the end of March and beginning of October, particularly to the south of inter continental convergence zone. Like the Indian south-west monsoon, these winds are associated with rainy weather and the season itself is rainy.

Gully erosion

Removal of soil by running water, with formation of channels, that cannot be smoothed out completely by normal cultivation.

Gully washer

A heavy rain that occurs suddenly.

Gust front

The leading edge of the cool, gusty surface winds produced by the downdrafts of thunderstorm.

Gust

A sudden and short positive departure from the ten minute average of wind speed. After a brief increase in the speed of the wind, what happens is that a lull follows and slackening in the wind speed takes place.

Gustnado

A weak, and short-lived, tornado that forms along the gust front of a thunderstorm.

Guttation

Exudation of water usually through structures called hydathodes present at the tips of veins of leaves. This occurs in plants growing under moist and humid conditions.

Gypsum block

Made up from plaster of paris with the help of two electrodes parallel to each other in a medium of gypsum block. It is a resistance unit which is used for measurement of soil moisture in 'situ' The resistance reading is a low i.e., 400-600 ohms at field capacity and high i.e., 50000-75000 at wilting point.

Gypsum requirement

The quantity of gypsum or its equivalent required to reduce the exchangeable-sodium percentage of a given alkali soil to an acceptable level.

Gypsum

A hydrated calcium sulphate ($CaSO_4$, $2H_2O$). The commercial material contains varying amounts of impurities and is used as a soil amendment.

H

Habitat

A specific kind of living space or environment where an organism lives or could be found or occurs.

Hadley cell

A mean thermally driven atmospheric circulation in which warm air rises and flows away from the equator at upper level and descends at the tropics. Wind moves from the tropics back to equator poleward. In many cases this term is also referred to as 'Hadley circulation'.

Hail

A form of precipitation. These are stone lumps of ice formed from protracted rimming in cumulonimbus updrafts. The size and shape vary a lot from a groundnut kernal to a cricket ball. Fundamentally, these are formed by strong vertical air currents carrying the rain water and by the accretion of ice crystals.

Hair hygrometer and hair hygrograph

These instruments are used to record the relative humidity of air. The hygrograph is useful for time recording of humidity of air in the atmosphere. In other words, a hygrograph is a recording hygrometer.

Halo

The ring of light that seems to encircle the sun or moon when veiled by cirrus clouds.

Hard pan

Is a hard and impermeable layer formed in the soil profile by accumulation of materials such as salts, clay etc., and which impedes drainage. Such hard pans are found in rice soils subjected for puddling years together.

Hard seed

Seeds that have a seed coat impervious to water or oxygen required for germination.

Hard water

Water which contains certain minerals, usually calcium and magnesium sulfates, chlorides or carbonates in solution, to the extent of causing a curd of precipitation rather than a lather, when soap is added.

Harmonic analysis

A point is said to be in simple harmonic motion when it oscillates along a line about a central point, so that its acceleration towards this point is always proportional to its distance from this point. When the periodic features are represented by addition of simple sine and cosine waves, then it is known as harmonic analysis. Here, it is proper to mention harmonic series which is a series in which reciprocals of the terms are in arithmetical progression.

Harrow

A secondary tillage implement that cuts the soil to a shallow depth for smoothening and pulverising the soil as well as to cut the weeds and to mix materials with the soil.

Harrowing

A secondary tillage operation which pulverizes, smoothens and packs the soil in seed bed preparation and/or controls weeds.

Harvest index (H.I)

- Yield of the plant parts of economic interest (economic yield) as percentage of total biological yield in terms of dry matter. It is given by the formulae

$$\text{H.I} = \frac{\text{Economic yield or grain yield}}{\text{Biological yield (Total dry matter)}}$$

- The economic yield of a crop divided by its biological yield.

Harvesting

The operation of cutting, picking, plucking or digging, or a combination of these for removing the useful part or economic part from the plants.

Haulm

Stem or stalk of a plant such as Cow pea, peas, groundnut etc. after harvesting.

Hay

Hay is feed produced by dehydrating green forage to a moisture content of 15% or less so that the biological processes do not proceed rapidly enough to build up heat.

Haze

This is largely considered as a form of condensation. The smoke, dust, pollutants, gases released from industries, etc., when get mixed up with water vapour, haze is formed, in which the visibility is reduced and the living organisms feel discomfort.

Head wind

A wind blowing in the opposite direction to the heading of an object in motion. The head wind is characterized as an opposition wind for the progress of an object. In micrometeorology, there is one thinking developing, of late, with reference to usage of this term in wind breaks of crop fields. This is used to refer to the wind in leeward side of a wind barrier which is dragged by the crop canopy because of which the viscal forces are reduced and the transfer of momentum reaches a balancing point.

Heat capacity

This is the heat required to raise the temperature of one cubic centimetre of a body by one degree centigrade (one degree Kelvin). In a highly compressible fluid like air, the value of heat capacity differs considerably depending upon the situation, i.e., whether at constant volume or pressure. For most of the soils, the heat capacity ranges between 0.3 to 0.6 calories per centimetre per centimetre. This is also known as 'volumetric specific heat'.

Heat balance

The balance between the radiation received by the earth (and atmosphere) from the sun and that redirected by the earth (and atmosphere).

Heat budget

The balance between the amount of incoming short-wave radiation received from the Sun and the outgoing long-wave radiation is known as heat budget of the Earth. It is genuine to mention a term 'Radiation balance' here, which is the net radiation flux and this is the balance between the net incoming short-wave solar radiation and the outgoing long-wave radiation at the surface of the Earth. Another term 'Energy balance' in crop canopies, means, the partition of net radiation into flux of heat into or out of the soil, sensible heat flux between the surface and air, the flux of latent heat to and from the surface through vaporization of water and the energy fixed in plants by photosynthesis.

Heat death

In this magnificent universe, the Earth is a minute part as compared to the space occupied by galaxies, stars, planets, comets, satellites, etc. It is to be noted that the temperature difference exists among various parts of the universe and so long as this situation continues it tries to attain thermal equilibrium. No doubt, it takes hundreds of millions of years. The reason is that the irreversible processes like radiation, conduction, etc., are continuously taking place and the entropy is going on increasing. Finally, the entropy tends towards maximum and the available energy towards minimum. A stage comes when all the heat energy will be at the same temperature and no energy will be available for use in converting into work. So, the universe unwinds itself. This is the heat death of the universe, and on some day it is inevitable and a reality.

Heat evasion

This is an opposite process of heat trapping. In tropics and sub-tropics, in certain seasons, heat load on plant canopies is above the tolerant limit. This situation results in excessive evapotranspiration. The microclimate also changes and the reproduction system of the plants will also be affected. However, by providing proper shading, it is possible to change the microclimate.

Heat exhaustion

The effect of excessive heat, on a living organism.

Heat index

The combined effect of air temperature and humidity.

Heat islands

Certain areas of the Earth are steadily warmer than the immediately surrounding areas. During nights, the cities are warmer than the surrounding rural areas. The variation in air temperature is more pronounced particularly in the central segments of the cities.

Heat Lightning

Lightning that appears as a glowing flash on the horizon from a distant thunderstorms.

Heat of vaporization

The quantity of heat energy required per gram to convert a substance from liquid to its vapour without change of temperature.

Heat stroke

The effect of high temperature on a body which is over exposed to heat

Heat transfer

The transfer or exchange of heat between the crop stand and its surroundings. Often this is also referred to as 'energy transfer'.

Heat trapping

This term is more a micrometeorological one as far as its applicability is concerned. There is ample scope to make this term itself into a technique. When the heat load on crop canopy is less than the required amount for physiological activities to be performed satisfactorily, the extra heat is generated by modifying the microclimate, by taking into account the angle of solar radiation relative to plants. Heat is trapped for compensating the low level heat in this way.

Heat units

Same as ;Growing degree days' and 'Effective heat units'.

Heat wave

A period of abnormally hot weather.

Heat

The heat contained by a body is the product of its mass, its temperature and its specific heat capacity. This is expressed in joules, calories, etc. The different definitions are as follows:-

The energy possessed by a substance in the form of kinetic energy of atomic or molecular translation, rotation or vibration.

The energy arising from random motion of all the molecules of a body.

The energy of a material stored in the form of sensible heat and/or latent heat.

The kinetic energy of molecular motion and the potential energy acquired on its expansion by a substance.

Heating (cooling) degree days

Same as 'Growing degree days' and 'Heat units'.

Heating degree day

This is counted for each degree that the daily mean temperature is lesser than a base temperature.

Heaving

This term refers to one of the damages caused by low temperature on plants, trees, etc. What exactly happens is, that, when a plant or a tree is growing, during winter, because of low temperatures they are either lifted upward or their roots stretch or break and sometimes the roots are pushed completely above the soil. This can be noticed on either side of the roads, while traveling in winter in northern States of our country.

Hekistotherm

A plant growing in a cold region. It requires less heat than other kinds of plants. As per the standard classifications, the average temperature for the warmest month of the year is less than $50°$ F for a hekistotherm to survive.

Helicity

A property of a moving fluid, representing the potential for helical flow (flow that follows a corkscrew pattern).

Heliophytes

Plants of sun loving species require intense light for normal development.

Heliotrophic

Inherent character of the plant species turning towards the sun ex: Sunflower

Heliotropic response

This is the response of some specific crops to the sunlight. For instance, the sunflower crop response to light is worth mentioning in this regard. As the Earth is rotating around itself and moving around the Sun in a cycle, the heads of sunflower plants also follow the same path of sunlight. In the morning a head faces east, towards the rising Sun. After following the path of the Sun, it faces west by evening and during night it rotates back to east and the cycle is completed and repeated. Such a response by a crop is known by this term.

Hemispherical reflectance of a surface

On any given surface, this is the ratio of total reflected flux to a radiant flux incident on that surface.

Henry

The unit of electrical inductance. In a closed circuit, the inductance is the rate of change of current of one ampere per second which produced an induced E.M.F. of one volt. The inductance may be self or mutual. Named after Joseph Henry (1797-1878),

Herbaceous plant

A vascular plant that does not develop woody tissue.

Herbage

Leaves, stems and other succulent parts of forage plants upon which animals feed.

Herbarium

A collection of plant specimens that have been taxonomically classified, pressed, dried and mounted on sheet of herbarium paper.

Herbicide

A chemical used for killing or inhibiting the growth of unwanted plants. They are also called as weedicides since they are targeted against weeds (unwanted plants).

Heterosphere

This is one of the atmospheric layers, the average height of which is 200 kilometres. In this portion of the atmosphere the gases are layered according to their molecular weight which clearly indicates that this is non uniform in physical and chemical properties. This is also called as thermosphere as the temperature increases with height.

Heterotroph

An organism that can not make its own food and hence obtains both organic and inorganic materials from the environment.

Heterotrophic

Capable of deriving energy for life processes only from the assimilation of organic carbon compounds and incapable of using carbon dioxide as the sole carbon source for cell synthesis. Contrast with autotrophic.

Hidden hunger

It is a term used to describe a plant that shows no obvious symptoms yet the nutrient content is not sufficient to give profitable yield.

High clouds

A term used to signify cirriform clouds that are composed of ice crystals and generally have bases above 7.5 Km and extending upto 12 Km.

High Latitudes

The latitude belt between 60° and 90° North and South.

High pressure system

An area of maximum pressure relative to adjacent air mass. It has diverging winds. Its rotation is opposite to the earth's rotation.

High pressure zone (Highs)

The regions of increased atmospheric pressure at mean sea level are usually referred to as 'high pressure zones' or 'highs' and more commonly as 'anticyclones'. Depending upon structure, storm tracks and general characteristics, these are again divided into different types. For example, the anticyclones developed in the sub-tropical regions are called as sub-tropical highs.

High sun

This term is used to indicate the season when the Sun is at its maximum noon altitude. This could be noticed in tropical countries particularly those that are located closer to the equator.

Hill agriculture

Farming systems practiced on high altitude terraced lands or valleys with rolling landscape.

Hill dropping

A method of sowing in which seeds are dropped at fixed spacing and not in a continuous stream. Thus the spacing between plant to plant in a row is constant.

Hill placement

This refers to applying fertiliser either in bands or near the plants. This method is usually employed when relatively small quantities are to be applied. Nitrogenous fertilisers are applied in pinches at the base of each plant of cotton, maize, cauliflower, cabbage, brinjal, tomato and other vegetable crops by this method.

Hingari

In Andhra Pradesh the late crop season from September to March.

Histogram

A bar diagram representing a frequency distribution.

Hoar frost

Frost that occurs when air with a dew point (below freezing) is brought to saturation by cooling.

Homosphere

On the basis of composition of gases by volume and weight the atmosphere is categorized as homosphere and heterosphere. The homosphere extends upto a height of 88 kilometres from the surface of the Earth. This layer is characterised by uniform composition of air. In certain cases, the depth is even up to 100 kilometres.

Hook echo

A radar reflectivity pattern observed in a thunderstorm. It appears like a fish hook.

Hook gauge

An instrument used to measure the changes in the level of water in the evaporation pans in a meteorological observatory.

Horizon

A plane or line used as reference for observation and measurement relative to a given location on the surface of the earth.

Horizontal distribution of temperature

The variation of air temperature depending upon the latitude is known as horizontal variation or distribution of temperature. The slow decrease in air temperatures poleward from equator is a classical example for this variation. On maps, isotherms are commonly used to show this distribution. The reason for this variation is the increasing slant of the Sun's rays resulting from the curvature of the Earth. The other reasons include differential heating of land and water, effect of ocean currents, mountain barriers, etc.

Horse latitudes

Located between 30° North and 30° South of the equator.

Horse power

The unit of power equivalent to 33,000 foot pounds per minute. One horse power is equal to 745.7 watts.

Hot air anemometer

An instrument used to measure the speed of wind.

Hub crop

Is the crop in a sequential cropping system which has the greatest comparative advantages over other crops e.g. vegetables/truck crops cultivated near city markets.

Hull (rice)

The rice caryopsis is surrounded by a hull (husk) composed of two modified leaves, the smaller palea and larger lemma.

Huller

A machine used for hulling. While removing the hull, the machine removes part of the bran also.

Hulling

It is the process of removing hull or husk and bran in one operation.

Humid mesothermal climate

A climate in which the annual temperature changes influence the type of vegetation. These climates are also called as warm temperate rainy climates. In many areas in these climates, wetness is found in all season. However, dry summers and winters are not uncommon. But winters are mild. On the Earth, these climates are restricted to the lower latitudes. Most of the vegetation in these climates survive moderate heat and moisture. The vegetation can not withstand low temperature winter seasons.

Humid zone

A climate characterized by hot and total absence of aridity. Well defined winter and summer seasons are seen. In sub-tropical regions, this zone is characterized by hot summers and mild winters. Occasionally, atmosphere contains adequate moisture.

Humidity mixing ratio

The weight of water vapour per unit weight of dry air with which it is mixed. The numerical values of mixing ratio differ significantly less from specific humidity in all situations except in total humid air. This is expressed in grams of water vapour per kilogram of dry air.

Humidity

The water vapour content of the atmosphere at any given point.

Humification

Process of organic decomposition leading to the formation of humus.

Humus

The well decomposed, more or less stable part of the organic matter of the soil.

Hurricane force wind

A tropical cyclone in the Indian Ocean is called as hurricane. These are the storms with a speed above 4 knots or 32 metres per second. On Beufort scale, the numbers are 12 through 1.

Hurricane warning

A formal advisory issued by forecasters when hurricane conditions are expected within a 24 - 36 hour period. The location, intensity, and movement etc., of the hurricane are informed in now-casting form.

Hurricane watch

A formal advisory issued by forecasters when the hurricane conditions are a potential threat within a 24 to 36 hour period.

Hurricane

A fierce tropical cyclone originates between 5 and 20 degrees north and south of the equator. Tropical cyclones are called as hurricanes in the North Atlantic and Eastern North Pacific Oceans. In the Indian Ocean these are called as 'cyclones'. These are found only in tropics and on oceans with a surface temperature of 80 degrees Fahrenheit, without any fronts. The 'eye' is calm with light winds but very strong winds upto 350 kilometres per hour blow around the central low pressure area. On Beufort scale, the wind force is 12 and above. In the Pacific, these are called 'typhoons'.

Hybrid vigour

The enhanced vigour of progeny resulting from the crossing of two different inbred plants.

Hybrid

The first generation progeny/offspring (F_1) from across between a male and female line.

Hydraulic conductivity

Is the proportionality factor 'K' in the Darcy's equation which states that the effective flow velocity 'V' in a porous medium is proportional to hydraulic gradient, (h1-h2)/L.

The practical application of hydrodynamics to engineering is hydraulics. In a flow of fluid, the effective flow velocity is proportional to the hydraulic gradient, which is known as hydraulic conductivity . In Darey's law this is the proportionally constant. It is to be noted that the fluids flow as a result of driving force. If a flow in a pipe is taken into consideration it occurs due to pressure differences across the ends of pipe.

Hydraulic equilibrium

Hydraulic equilibrium is the condition for zero flow rate of water film in soil.

Hydraulic head

Is the elevation with respect of a standard datum at which water stands in a riser or a manometer connected to a point in question in the soil.

The summation of fluid pressure head and elevation with respect to a given situation in an experiment or an observation point.

Hydraulic radius

Is the ratio of the volume to the surface of the pore space or the average ratio of the cross sectional area of the pores to their circumferences.

Hydrography

The science that deals with the waters of the earths surface, particularly with reference to their physical features, position, volume etc., and the precipitation charts of seas, lakes, rivers, contours of the sea bed, shallows, deep currents etc.

The study of water for its origin, behaviour and other properties in the atmosphere is commonly referred to as hydrology. Such water flows into tanks, rivers, seas, etc. The overall phenomena in the movement of fluid water, the description of flow and other characteristics is known as hydrography.

Hydrological cycle

The hydrosphere of the Earth is two parts out of the total three parts of the Earth. The state of water change from one to another as a pathway, continuously. The changes start as evaporation from the soil and finally reach back to soil as water after going through the phases of condensation and precipitation. The total change of water as a series through the water bodies on the Earth is known as hydrological cycle.

Hydrological drought

This is a situation in which hydrological resources like streams, rivers, reservoirs, lakes, wells, etc., dry up because of marked depletion of surface water. The ground water and other income generating major sources are affected. If meteorological drought is significantly prolonged hydrological drought sets in.

Hydrology

A branch of engineering science, which deals with the study of the total properties of water both on the earth and in the atmosphere.

The science that deals with water specially in relation to its occurrence in wells, lakes, streams etc., and as snow including its uses, conservation, control and discovery.

Hydrolysis

Breakdown of substance with the addition of water.

Hydrometer

An instrument used to find the specific gravity of a liquid.

An instrument used to measure the density or specific gravity of liquids.

Hydrophytes

Plants growing habitually in water or in very wet soils where at least periodically oxygen becomes deficient as a result of excessive water content.

Hydroponics

- The growing of plants in water solutions of essential nutrients. It is other wise called as soil less culture.
- The growing of plants so that the roots are immersed in a water solution of nutrient salts or in some inert material such as vermiculite which is supplied with a nutrient solution. This system is practiced, when there is scarcity for land.

Hydrosphere

The water portion of the Earth is known as hydrosphere. This water portion includes not only oceans but also inland water bodies. Of late, some scientists believe that the ground water and atmospheric water substance are also part of hydrosphere.

Hydrostatic equation

The recorded equation of hydrostatic equilibrium. See 'Hydrostatic equilibrium'.

Hydrostatic equilibrium

A state of balance between opposing forces and effects is known as equilibrium. A state of equilibrium applies very accurately to many parts of the atmosphere. But, in case of a disturbed part of the atmosphere, it is not possible. The balance between the gravitational downward pull on an air parcel and upward pressure gradient force arising from vertical pressure lapse is referred to as hydrostatic equilibrium, and this applies to an undisturbed part of the atmosphere.

Hydrostatic pressure

Fluids existing in different states are subjected to pressure because of several reasons. Actually, the study of forces and pressures in liquids, based on mathematics is hydrostatics. So, when a pressure in a fluid is exclusively due to the weight of the fluid above, then the same is called as hydrostatic pressure.

Hydrostatic stability

In the atmosphere this stability is noticed when hydrostatic balance exists between the vertical pressure force and gravitational force. It is to be noted that a column of air is said to be stable when its existing lapse rate is less than the dry adiabatic lapse rate.

Hydrotropism

The growth movement in response to unequal distribution of water, as when roots bend towards moist soil.

Hyetal

Belonging to rain.

Hygrometer

An instrument used to measure the water vapour content of the atmospheric air 'in situ'.

Hygroscope

An instrument used to measure the variations of relative humidity of the air. A hygroscope readily takes up and retains the moisture.

Hygroscopic water

- Is that amount of water which is absorbed by the soil from an atmosphere of saturated water vapour due to attractive forces on the surface of the particles.
- Water held so firmly by the attraction of soil particles that it can be removed only by heating above 100 ^{o}C . It is not available for plants to perform its activities and processes. The thin film of water which is present on the soil particles and not available to plants for its routine mechanisms.

Hygroscopic

This term is commonly used to refer to small particles in the atmosphere which possess the tendency to attract, absorb and condense the ambient water vapour.

Hygroscopicity of fertiliser materials

Some fertiliser materials absorb moisture from the air, which causes them to become sticky, cacking such hygroscopic fertilisers are calcium ammonium nitrate, ammonium chloride and urea.

Hygrothermograph

An instrument used to measure both the temperature and humidity of the ambient air. This is a combination of both thermograph and barograph.

Hygrothermoscope

An instrument used to measure not only humidity and temperature but also the variations in dew point.

Hypogeal germination

A type of germination in which the cotyledons remain below the soil surface.

Hypothermia

The failure of the body to maintain adequate production of heat under conditions of extreme cold conditions.

Hypothesis

- A speculative statement based on observation suggesting a possible cause and effect relationship and forming the beginning point of an experimental study.
- The proposed law put forward in explanation of observed facts. The hypothesis must be reasonable to subject the same to the experimental test.

Hypsometer

An instrument used to measure the altitudes of various points. A specific feature of the instrument is that the use of the boiling point of either water or any other liquid for determining the atmospheric pressure and the *barometric formula* is used to find the difference in elevation.

Hysteresis

Lag in one of the two associated processes or phenomena.

Hyther graphs

The plots of precipitation and temperature or humidity and temperature. These plots are for different periods but commonly monthly plots are seen in the field of applied physical sciences.

I

Ice age

Of different theories of formation of the Earth, a common belief is that the Earth had passed through certain ages in general and with reference to the total area in particular. As per the view point of meteorologists and geologists the ice age is considered as the time when the cryosphere of the Earth is approximately twice the present area.

Ice crystals

Water in solid state is called ice. If this solid form is without any branches and the shapes are like needles, columns or plates and if their fall is very slow, then the situation is described as falling of ice crystals.

Ice fog

Fog that is composed of minute ice particles.

Ice jam

The accumulation of broken ice caught in a narrow channel. This may produce local flooding.

Icelandic low

A semi-permanent, subpolar area of low pressure in the North Atlantic Ocean.

Ice pellets

Precipitation in the form of transparent or translucent pellets of Ice. With a diameter of 5 mm or less.

Ice storm

A severe weather condition characterized by falling of ice in heaps.

Icicle

Ice that forms in the shape of a narrow cone hanging point down.

Icing

process of deposition of ice on an object.

Ideotype

Refers to plant type in which morphological and physiological characteristics are ideally suited to achieve high production potential and yield reliability. This term was first coined by 'Donald' while working on wheat crop improvement. Ideotypes otherwise called as "New Plant types".

Illumination value

The illumination capacity or brightness of light as perceived by the human eye.

Illumination

The amount of light falling on a unit area of the surface per second is known as illumination of a surface. So, this is a general term to describe how well a surface is lighted. Another term illuminance is that measurable quantity to describe the amount of luminous flux received by one unit area of a surface.

Illuviation

The process of deposition of soil material removed from one horizon to another horizon of the soil usually from an upper horizon to a lower horizon in the profile.

Imbibition

The uptake of a liquid with swelling by a substance such as seed, cellulose, agar or gelatin.

Immature soil

A soil with indistinct or only slightly developed horizons because of the relatively short time it has been subjected to the various soil-forming processes. A soil that has not reached equilibrium with its environment.

Immobilization

The conversion of an element from inorganic to organic combination in microbial or plant tissues. This has the effect of rendering unavailable (and usually not readily soluble) an element that previously was directly available to plants.

Impedance

This is the property of a wave conducting media which expresses the ratio between the rate at which it transmits the energy and the amplitude of the wave. In the alternating current circuit this is the quantity which determines the amplitude of the current for a given voltage.

Impermeable (seed)

Impermeable seed is one whose seed coat allows no passage through it water or gases.

Impervious soil

A soil through which air, water or plant roots penetrate very slowly if at all.

Implements

The equipment used on farm to carry out different farm operations either with the help of tractor or bullock power.

Importance value (I.V.)

The importance value of a species indicates the degree of dominance of a species over the other species in a given sample plot.

Impoverished soil

Soil which has been continuously cropped and has had little or no care so that it is incapable of supporting a satisfactory or economic crop.

Improved production technology

Is the use of improved varieties and associated improve crop management practices followed on a farm.

Improved seed

The genetically and physically pure seed of an improved crop variety. Its different categories are nucleus seed, breeder's stock seed, foundation seed, registered seed and certified seed.

Impulse

This is the product of the magnitude of the force and the time during which the force acts. This is equal to the amount by which the momentum changes or also equal to the total change of momentum produced by it.

In situ

In the natural or original position or location.

Inbred

The progeny of either a single cross-pollination plant obtained by selfing or two closely related plants obtained by inbreeding.

Inches of mercury (Hg)

The height of a column of mercury in a barometer.

Incipient wilting

Small loss in turgidity during warm weather, even when soil is moist. It may not lead to any visible stress symptoms like rolling or folding of leaves and is usually reversible at night.

Inclination of the wind

At any particular location, the angle between the direction of the wind and the gradient wind.

Income equivalent ratio (IER)

The ratio of the area needed under sole cropping to produce the same gross income as one hectare of intercropping at the same management level.

IER is the conversion of LER into economic terms, thus

$$\text{IER} = \text{value of combined inter crop yield} \times \frac{\text{LER} - 1}{\text{LER}}$$

where, LER = Land Equivalent Ratio.

Incompatibility

Inability of the gametes of two organisms to unite and reproduce sexually. Inability of one substance to mix with other without undesirable reaction e.g. with references to insecticides, herbicides, fungicides where incompatibility involves loss of efficiency on mixing.

Incorporating or mixing

Tillage operations which mix or disperse foreign materials, such as chemicals or plant residues into the soil.

Incremental cost benefit ratio (ICBR)

A ratio of benefit to cost invested for obtaining economic returns in rupees per rupee further invested for the costlier, valuable inputs imposed, such as fertilizers, weedicides, pesticides, fungicides etc., for achieving the optimum maximum yields of a crop. ICBR is always compared with the next lower level of application of the respective input.

Independent and dependent variables

The symbol which denotes a member of the domain of a function is called an independent variable, and the symbol denoting a member of the range of a function is called a dependent variable.

Indeterminate growth

The flowers are borne on lateral branches, the central stem continuing vegetative growth with blooming continued for a long period like in cotton or Redgram.

Indeterminate plants

Plants which flower when the days are either long or short.

Index cycle

In mathematics, index is the number indicating the power to which the quantity is raised. In meteorology, the same word is used to indicate the pressure difference between the latitude zones. However, to express the frequency of alteration between periods of zonal and meridional flows, 'cycle' is added to 'index', to emphasize precisely in a more scientific manner.

Index of combustibility

An index which is useful in forecasting the fire weather. The main parameters which are generally involved in arriving at the index are the air temperature, S.V.P. deficit, number of days since last rainfall, etc.

Index of refraction

When a ray of light travels obliquely from one medium to another, it is bent or refracted at the surface separating the two media. So , the index of refraction is the ratio of velocity of light in vacuum to its velocity in a given medium.

Indian standard time (IST)

The surface of the Earth is divided into 24 'time zones', the way in which there are 24 hours in a day. The time established in each of the zone is called 'standard time'. The Indian standard time (IST) is the local mean time (LMT) for the meridian of longitude 82° 30' E. This is the longitude of Allahabad which is taken as standard longitude for India. Since each degree is equal to four minutes of time, the IST is five and a half hours ahead of Greenwich mean time (GMT). The GMT is also known as universal time.

Indian summer

A period of abnormally warm weather from March to May with clear skies and cool nights.

Indicator plant

A plant which reflects either by its presence or character of growth, specific growing conditions like deficiency of plant nutrient, soil moisture stress etc. Example sunflower is taken as indicator for irrigation and maize is taken as indicator for zinc deficiency.

Indigenous

An organism that is native to a particular habitat, as distinct from one introduced from out side the area.

Inertia

This is the tendency of a body to preserve its state of rest. In other words this is the resistance and opposition offered by a body to any attempt to change its direction or rate of motion.

Inertial reference frame, the tendency of a body to preserve its state of rest or uniform motion in a straight line is called as 'inertia'. A reference frame with zero absolute temperature is known as Inertial reference frame.

Infiltration capacity or soil infiltrability

Is the flux which the soil profile can absorb through its surface where it is maintained in contact with water at atmospheric pressure and there is no divergent flow at the borders.

Infiltration rate

Is the volume of water passing into the soil per unit area per unit time.

Infiltration

Is the process of water entry into the soil generally (but not necessarily) through the soil surface and vertically downward. Water may enter the soil through the entire surface uniformly as under ponding or rain or it may enter the soil through furrows or crevices, or it may move up into the soil from a surface below under high water table.

Infrared radiation

The solar radiation is higher than visible wavelength segment with wavelengths between 0 to 50 microns. This is popularly known as infrared radiation, because this is the electromagnetic radiation being behind the red end of the visible spectrum but not far behind.

Infrared thermometer

This is a non-contact thermometer used to measure the temperature of the crop plants. Infrared energy is emitted by all solid objects above absolute zero. The amount of energy emitted is proportional to the body or target temperature. The infrared sensor collects this energy by means of fixed focus optics onto a sensitive detector and reads out directly in ^{0}F or ^{0}C. It is fast because it collects the IR energy at the speed of light, and the detector has very low mass. The time constant is 0.1 second, about 10 times faster than conventional contact methods.

Initial intake rate

Rate at which the water enter the soil when first applied and it is measured in mm / hr.

Initialization

The micrometerological data shall be put in order before attempting any physical principle comprehensively to arrive at reasonable and valid conclusions. Initialization is the first step in preparing the data for use as input either in a computer or any other system of processing.

Inoculate

To bring the pathogen into contact with the host, organ or medium.

Insolation

The incoming solar radiation from the Sun that strikes the surface of the earth over a finite time period such as a day.

Instability

The condition of a system, which responds to a specific disturbance by increasing the disturbance until an irreversible change has taken place. Another term 'Conditional instability' refers to the state of a column of air when its vertical distribution of temperature is such that the layer is stable for dry air but unstable for saturated air.

Instrument flight rules (IFR)

The general weather conditions that pilots are briefed at the surface.

Instrument shelter

A wooden or steel base designed to protect the weather measuring instruments from exposure to direct sunshine, precipitation, ventilation etc.

Integrated weed management index (IWMI)

It is the average value of weed management Index (WMI) and Agronomic management Index (AMI). The formula is

$$IWMI = \frac{WMI + AMI}{2}.$$

Integrated pest control

Uses a variety of technologies in a single pest management with the objective to produce optimum crop yield at a minimum cost taking in to consideration ecological and socio-economic constraints under a given agro-ecosystem.

Intensity of the radiation

The power emitted or received from a source of radiation per unit solid angle. In other words, this is the flux radiated per unit solid angle from a point source. This has a directive property and the units are watts per steridian.

Intensity

This word is usually referred to denote strength. For instance, the strength of an electric field at a point is known as electric intensity. In the same way, radiation intensity.

Intensive cropping

Maximum use of the land by means of frequent succession of harvested crops.

Intensive property

A property that is independent of the size of a sample, e.g., latent heat, sensible heat, specific heat, etc., of either air or soil.

Inter glacial

'Inter' is a prefix denoting between or among and 'glacial' is an usual term referred to ice age. So, the period between consecutive ice ages in succession is inter glacial period.

Inter tilled crop

A crop planted in rows followed by cultivation between the rows.

Inter tropical convergence zone (ITCZ)

A broad zone in low troposphere which is confined to low latitudes only. Near the equator, ascent of air results in low atmospheric pressure, deep convective clouds, the consequence of which is heavy precipitation. Because of the movement of the Earth and also due to change in solar radiation received, the position of ITCZ also changes seasonally.

Interaction effect

Measured by the failure of simple effect of one factor to remain consistent over the various levels of the other factor. The change in response of one factor in the presence of the other factor.

Intercropping

Refers to growing of two or more generally dissimilar crops simultaneously on the same piece of land. Base crop necessarily in distinct row arrangement. The recommended optimum plant population of the base crop is suitably combined with appropriate additional plant density of the associated crop, and there is crop intensification in both time and space dimensions.

Intercrops

The crops raised in an orchard or other widely spaced crops for increasing the income from the same piece of land e.g., short duration vegetables, pulses, oilseeds etc.

Intercultivation (interculturing)

Soil cultivation performed in between the standing rows of the crop.

Interface

A boundary which separates two fluids of different properties.

Interference

This term is usually referred in case of wave motions. If the crest of one wave meets the trough of another of equal amplitude then the wave is destroyed at that point. Incidentally, the superposition of one crest upon another leads to an increased effect. So, with this information it can be stated that the mutual effect of two waves of light that arrive at the same time causes interference.

Intermountain high

An area of high pressure that occurs between the mountains.

Internal drainage of soil

The quality of soil that permits downward flow of excess water through it. This is determined by the texture, structure, depth of water table, etc., of a cropped site. In crops like rice this is a major factor which determine its water requirements.

Internal energy

This is the sensible heat energy in an ideal gas at constant volume. This is assessed from the specific heat capacity of that gas, also, at constant volume. In micrometeorological studies of crop plants, the heat flux which is utilized by the air inside a crop canopy is taken as internal energy.

Internal environment

The conditions within an organism or cell that influence its processes, e.g., air spaces in plants. Often, sensible heat flux influence the internal environment of a crop.

International date line

The line of longitude located at 180° east or west (of the prime meridian) where the date changes by a day. West of the line it is one day later than east of the line.

Interplanting

In woodland, the setting out of young trees among existing trees or bushy growth. In orchards, the planting of the farm crops along the trees, especially when trees are too small to occupy the land completely. In crop land, the planting of several crops together on the same land like the planting of soybean with maize.

Inverse nitrogen yield concept

According to Wilcox, the yield of the crop is inversely proportional to its nitrogen content i.e., Grain yields is $\alpha \dfrac{1}{N}$, where, 'N' = nitrogen content in dry matter.

Inversion

A situation in which there is an increase of atmospheric temperature with height instead of fall at the same altitude. What exactly happens is that in the lower part of the atmosphere upto a height of 8 to 18 kilometres from the surface, the temperature normally decreases at the rate of 0.5 degrees per each kilometer. But sometimes under special circumstances, it is reversed in certain layers of the atmosphere. In other words, the vertical pressure gradient is inverted.

Ions

A charged atom or particle formed usually by the dissociation of a molecule.

If an electron is removed from an atom then it is said to be ionized. Ions are charged particles in a fluid. It is to be noted that the ions may be formed either by gaining or by losing the electrons, in a gas or of any other matter.

Ionisation of gases

The process of formation of ions from gas atoms is known as ionization of gases.

Ionizing radiation

The alpha, beta and gamma radiations are called ionizing radiations, because they take electrons from atoms and attaches them to other atoms. In the atmosphere, the ionosphere exhibits this character.

Ionosphere

This is the layer of the atmosphere which lies beyond the ozonosphere at a height of 80 km above the surface of the Earth. At 80 km, the temperature again falls to −100 degrees centigrade and upto 400 km this layer consists of charged particles. This layer is always chemically active because of the presence of electrically charged ions.

Irrigated dryland

Areas with welldrained soils where water does not accumulate and irrigation is used mainly for dryland crops such as wheat, maize, potato, and other cash crops. Dryland rice may also be grown with intermittent type irrigation.

Irradiance

The radiant flux density incident on a unit area of a real or imaginary surface.

Irrigation efficiency

Is the ratio expressed in percentage of water stored in the root zone depth of the soil to the water delivered in the field from the farm supply source.

Irrigation interval

Time between the start of successive field irrigation application on the same field, in days.

Irrigation period

Is the number of days that can be allowed for applying one irrigation to a given design area during the peak consumptive use period of the crop being irrigated. It is the basis for designing the capacity of an irrigation project.

Irrigation requirement

Is a part of total water requirement of a crop, exclusive of effective rainfall and soil moisture stored in the root zone or that contributed from the shallow ground water table. i.e.,

$$IR = WR - ER - GW_c - Ws_r$$

where

IR = Irrigation requirement

WR = Water requirement

ER = Effective rainfall

GW_c = Ground water contribution

Ws_r = Water stored in root zone

Irrigation response

Is the rate of increase in crop yield per unit of increase in water applied.

Irrigation water requirement

The net irrigation water requirement divided by the irrigation efficiency i.e., $\dfrac{\text{Net irrigation}}{\text{Irrigation efficiency in \% (x\%)}} = \dfrac{\text{Net irrigation}}{x} \times 100$

where

x = Irrigation efficiency that may be 60% or 70% depending upon the soil and condition of layout.

Irrigation

The application of water to soil to assist in the production of crops, especially during stress periods.

Isallobar

The line of equal change in atmospheric pressure during a specific observation time period.

Iso

A prefix denoting 'equal'. For instance, isotherms, isobars, etc., are the lines drawn through place of equal temperature and pressure respectively.

Isobars

The imaginary lines drawn on a weather map connecting points of equal atmospheric pressure.

Isobath (water-table)

A line on a map or ground connecting points at the same height or elevation above an aquifer or water-table.

A line drawn on a map connecting points of equal depth at the bottom of the sea.

Isodrosotherm

The line drawn on a weather map connecting points of equal dew point.

Isohel

A line or curve formed by joining the points of equal sunshine duration over a given period of time.

Isohytes

The lines drawn on a weather map, joining places that include districts which experience same rainfall, for a given period of time.

Isolated system

A system in which the energy is neither lost nor gained from outside. If one considers a system in which the particles can exchange energy and interact with one another, only then the system can be called as an isolated system.

Isopleth

A line with uniform value of a given scaler quantity like isobar, isotherm, etc.

Isopycnal

A vast isopleth of air density.

Isotach

An isopleth of wind speed.

Isothermal change

When a change in pressure and volume of a substance takes place, but the temperature remains constant, then the change is said to be isothermal. For instance, the isothermal expansion of a gas is an isothermal change because this change takes place at constant temperature.

Isothermal layer

A layer in the atmosphere with constant temperature, i.e., for a given change in the height there is no change of temperature. This is more an imaginary situation except under zero lapse rate conditions.

Isotherms

The lines on a weather map joining places of equal temperature on the Earth for a given period of time.

Isotopic fractionation

Atoms of the same element which differ in mass number are called the isotopes of the same element. If separation of such isotopes occurs, then it is said to be fractionation of isotopes. It is to be underlined that isotopes of atoms possess different number of neutrons.

Isotropic radiation

This term shall not be mistaken for yet another type of radiation. The term isotropic means, exhibiting uniform properties throughout and also in all directions. If the solar radiation received at a particular height in a crop canopy is diffused in many directions and the intensity of the radiation is also same, at all levels in these directions, then the radiation is called as isotropic radiation.

Isotropic

An atmospheric phenomenon which is independent of direction. For instance, the mesoscale turbulence is nearly isotropic. It does not take into consideration the length of growing period for various crops.

IW/CPE ratio

Based on the climatological approach, water is applied to the field and the depth of irrigation water is fixed for different crops i.e.,

$$\frac{IW}{CPE} = ratio$$

where, IW = Irrigation water in depth (cm), CPE = cumulative pan evaporation (cm), ratio = a numerical value usually 0.8, 0.9 or 1.0.

It is one of the criteria for scheduling of irrigation to field crops.

J

Jet streak

A region of accelerated wind speed along the axis of a jet stream.

Jet stream

A strong and narrow stream of flowing air. In the high troposphere, the flows are at 300 to 400 kilometres per hour. In mid latitudes, a jet stream is the 'core' of planetary waves.

Jhum cultivation

The slash-and-burn type of shifting cultivation in the hill tracts of Bangladesh and Assam.

Joule

The unit of work or energy. In mechanical systems, this is the work done when the point of application of one Newton is displaced at a distance of one metre in the direction of force. In electrical systems, a joule is the equivalent of moving one coulomb through a potential difference of one volt. One joule is equal to one watt per second.

Joule's law

For an ideal gas, the internal energy at constant temperature is independent of the pressure and volume.

Joule-Thomson effect

When a gas is allowed to expand through a narrow orifice, cooling is produced. This cooling is proportional to the pressure differences across the orifice.

Julian day

The number of a day in an year, starting from January 1st for counting. January 1st is the day one and December 31st is the last day of the year. For example, February 5th is Julian day number 36 and February 10th is Julian day number 41.

Juvenile phase

The phase of vegetative growth of a plant before which it can not be induced to produce flowers. This period may vary from plant to plant e.g. plants have a juvenile phase of one month.

Juvenile water

Water that has been trapped for ages far below the Earth's surface, so that it cannot take part in the hydrologic cycle.

K

K index

The measure of thunderstorm potential based on the vertical temperature lapse rate, the moisture content and extent vertically.

Katabatic wind

This is a valley-wind of down slope drainage type, developed mostly on clear nights when regional pressure gradients are weak. This is caused by surface level cooling of the atmosphere and the wind blows down an incline. If warm it is a 'foehn', if cold it may be a 'fall' or a 'gravity wind'.

Katafront

A front where the warm air descends the frontal surface.

Kelvin temperature scale

An absolute temperature scale in which the zero point is absolute zero. The scale is based upon the triple point of pure water and the size of the degree is same as on the Celsius scale. So the 273.16 ^{o}K = 0 ^{o}C. That means the freezing point of water is 273.16 ^{o}K = 0 ^{o}C, etc. This is more commonly called as the absolute scale.

Kelvin

The unit of absolute temperature.

Kew pattern barometer

A fixed scale mercurial barometer. In this barometer, only one adjustment is needed before each recording.

Kharif crops

Crops grown during the main monsoon season (June to September), such as kharif sorghum, maize, bajra, rice, cotton etc.

Kharif

In Andhra Pradesh the early crop season from June to December.

Kilo calorie

Commonly used term in diets as 'calorie', meaning a thousand calories.

Kilogram

Kilo-a prefix meaning 1000 of. Hence, a kilogram is 1000 grams and so is the kilometer. Kilogram is the unit of mass defined as the mass of a standard metal cylinder.

Kinetic energy

The energy that a body possesses in respect of its motion. It is given by half the product of the mass and the square of the linear velocity of the body.

Kinetic theory

The theory that has been highly successful in explaining the behaviour of gases on the basis of the ordinary laws of mechanics, applied statistically to the motions of the molecules. The main feature are (a) All gases are composed of molecules. (b) The molecules are point masses and are continuously in motion. (c) The molecules continuously collide against each other in a steady state and the distance moved in a straight line is called 'freepath' of the molecules. The collisions are instantaneous. (d) In a direct impact the molecules begin to move in the opposite direction with the same velocity. (e) On an average the number of molecules remain constant per cubic centimetre.

Kirchoff's law

Absorptivity of an object for radiation of a specific wavelength is equal to its emissivity for the same wavelength.

Knot

Wind speed is measured in several units such as metres per second, kilometers per hour, etc. A knot is also another unit, which means one nautical mile per hour.

Koppen system

A very popular and standard climatic classification developed by Koppen.

Koppen's classification of climates

The classification of climate devised by Wladmit Koppen, a German botanist and climatologist, is both quantitative and empirical. In this classification, the extremes of temperature, extremes of precipitation, etc., are characterized by an index composed of a certain number of letters. In turn, these will give an immediate summary of the principle characters of the various climatic elements.

Kotka phosphate

A product obtained by partial acidulation of ground rock phosphate with sulphuric acid and containing not less than 16 per cent of available P_2O_5 of which at least 6 per cent is water soluble.

Kreb's cycle

The aerobic portion of respiration, in which pyruvic acid is oxidized, usually to carbon dioxide and water as end products. This is also known as TCA cycle or citric acid cycle.

L

Label

All written, printed or graphic matter on or attached to the economic poision, or the immediate container thereof, and the outside container or wrapper to the retail package of the economic poison.

Lake effect snow

The showers of snow that are created when cold dry air passes over a large warmer lake.

Lambert's cosine law

The law states that, when radiation is emitted by a full radiator at an angle (angle of incidence) to the normal, the flux per unit solid angle emitted by unit surface is proportional to cos of the angle of incidence. The law states that, the radiant intensity (flux per unit solid angle) emitted in any direction from a unit radiating surface varies as the cosine of the angle between normal to the surface and the direction of radiation.

Laminar flow

The flow in which adjacent volumes of fluid move in a regular manner with respect to each other. In laminar flow viscosity dominates and fluid moves in streamlines without any turbulence.

Laminar sub layer

Laminar means thin sheet like. If the vertical motion of eddies is not present in a relatively small layer of a fluid, then the layer is said to be laminar sub-layer.

Land (marginal)

Land from which it is doubtful whether a specified form of management can produce an income in excess of the costs of such management under given economic conditions, usually with reference to agricultural land.

Land breeze

A breeze from the land to the sea due to the difference in temperature between land and water. Basically, this is a surface offshore flow at night. In general, at night, land temperature falls rapidly as compared to water surface which results in cooling of air over the land. A pressure gradient is thus developed between the cool air over the land and warm air over the sea. This results, finally, as a surface offshore flow at night. In a nutshell, this is the counterpart of sea breeze.

Land capability

The suitability of land for use without damage. It is an expression of the effect of physical land conditions on the total suitability of land for different uses without damage for crops that require regular tillage, for grazing, for woodland, for wildlife and other recreations.

Land characteristic

An attribute of land that can be measured of estimated, and which can be employed as a means of describing land qualities of distinguishing between land units of differing suitabilities for use.

Land clearing

Removal of heavy vegetative growth from the land prior to land preparation.

Land drain

A drain drawing of water from land.

Land equivalent ratio (LER)

It denotes the relative land area under sole crop required to produce the same yield as obtained under a mixed or an intercropping system at the same management level. It is calculated as sum total of the ratios of yield of each component crop in an intercropping or a mixed cropping system to its corresponding yield when grown as a sole crop

thus: $\text{LER} = \sum_{i=1}^{m} \dfrac{y_{ij}}{y_{ii}}$

where,

y_{ij} = Yield of i - th component in inter cropping system.

y_{ii} = yield of i - th component grown as sole crop over the same area.

Land evaluation

The process of assessment of land performance when used for specified purpose, involving the execution and interpretation of surveys and studies of landforms, soils, vegetation, climate and other aspects of land in order to identify and make a comparison of promising kinds of land use in terms applicable to the objectives of the evaluation.

Land facet

A land unit (q.u) with climate, landforms, soils and vegetation characteristics which for most practical purposes may be considered as uniform. A subdivision of a land system.

Land fall

The point at which the eye of a tropical cyclone first crosses a land mass.

Land forming

Tillage operation which move soil to create desired soil configurations. Forming may be done on a large scale such as contouring or terracing or on a small scale, such as ridging or pitting.

Land grading

Reshaping of the field surface to a desired elevation and slope. It is necessary in making a suitable field surface to control the flow of water, to check soil erosion and to provide surface drainage.

Land levelling

The reshaping of the land surface to facilitate a more uniform application of irrigation water or for preventing soil erosion. The perfect land levelling can be achieved by laser beams.

Land planning

Tillage operation that cuts and moves small layers of soil to provide a smooth and refined surface.

Land reclamation

- Making land capable of more intensive use by changing its character and/or environment through such operations as land clearing, controlling erosion, construction of irrigation reservoirs etc.

Land side

That part of the plough that sides along the face of the furrow wall. It helps to slice on the ground and also helps in stabilizing the plough while it is in operation. Land side is fastened to the frog, another part of the plough with the help of bolts.

Land slide

- Rapid movement downslope of a mass of soil, rock and debris.

Land spout

A weak tornado beneath cumulonimbus (or) to wearing cumulus clouds which is the land equivalent of a water spout.

Land suitability

The fitness of a given type of land for a specified kind of land use.

Land use patterns

Alternative ways to utilize available land resources over the time for agricultural production.

Land utilization index

Has been defined as the number of days during which the crops occupy the land during a year divided by 365 and can be expressed as a fraction or as a percentage.

Land

That part of the surface of lithosphere which is not usually covered with water, forms the total natural and cultural environment with which production takes place.

Langley

A unit of radiant energy. This is equal to one gram calorie per centimetre per centimetre.

Lapse rate

In meteorology, the rate of decrease of temperature with increasing height, at a given time and place. Under normal conditions, a negative temperature gradient is a positive lapse.

Lapse

Same as lapse rate See 'Lapse rate'.

Latent heat flux

- For common usage and general understanding 'flux' is some substance added to the already existing substance so that the same will assist in generating energy. The latent heat flux is the amount of heat energy utilized in evaporation and transpiration in a crop canopy. This energy is also a part of the solar radiation passing through the canopy.

Latent heat

The heat energy either released or absorbed per unit mass in a reversible, isobaric and isothermal change of phase. In this process, the 'mass' is either a gas or a liquid like water when either liquify, solidify, melt or evaporate etc., in which, a change of phase takes place.

Lateral acceleration

Acceleration across the direction of flow. This may be seen in different types of cyclones, wind systems, flow of fluids, etc.

Latitude

The imaginary circular lines drawn horizontally connecting East and West on the globe. These lines help to find out the distance of a place from the equator and are expressed in degrees. The equator is imagined as zero and the poles as ninety degree latitudes.

Law of additive probability

It states that if two events A and B are mutually exclusive, the probability of occurrence of either A or B is the sum of individual probability of A and B.

Law of average product

Average product is the total output divided by the total input.

Law of constant return

Constant productivity holds true if all units of the variable factor which are applied to the fixed factor, result in equal additions to the total output.

Law of Dalton

When a mixture of several gases are maintained over water, each dissolves independently and in accordance with the gaseous pressure (partial pressure) which it exerts against the surface of the water.

Law of diminishing return

Diminishing productivity of the variable factor exists when each additional units of inputs adds less to output than the previous unit.

Law of frequency

The generalization which states that, when frequency indices of species in a stand are classified into five main classes a double peak occurs in homogenous vegetation.

Law of Henry

The mass of a slightly soluble gas that dissolves in a definite mass of a liquid at a given temperature is a very nearly directly proportional to the partial pressure of that gas.

Law of minimum

States that when a character such as yield is controlled by number of separate factors, the rate of the process is limited by the factor which is present in relative minimum.

Lay out

The placement of experimental treatment on the experimental site where it may be over space, time or type of material.

LD50 (pesticide)

Abbreviation of median lethal dose. It indicates the amount of toxicant necessary to effect a 50% kill of the pest being tested. Measured in mg/kg of body weight.

Leaching

The removal of soluble constituents from the soil by percolating water. In paddy crop, when it is grown under puddle and flooded conditions unless an impervious layer is formed by puddling process, there are leaching losses to a greater extent.

Leaf area duration (LAD)

Is a measure of the ability of the plant to produce and maintain leaf area and is obtained by integrating the leaf area index over crop growth period. It is usually expressed in days or weeks. LAD = LAI × time interval, where LAI = Leaf area index.

Leeward

The direction towards which the wind blows.

Leewave

A stationary disturbance of wave in the airflow over an obstacle or a barrier. Here, the barrier may be a hill, a mountain or a range of hills and the disturbance is usually due to buoyancy.

Legume

A dry dehiscent one-celled fruit which splits along the suture of the single carple a pod. Example pea, bean. A plant having such fruits is called legume.

Lenticular cloud

Anything pertaining to a lens, particularly a bi-convex lens is termed as lenticular. A lenticular cloud is the one which is having a sharp edge and lens in shape. If a large number of such clouds are seen in a given situation it looks like a pile of plates staked in a series.

Lenticular transpiration

The openings present on the older stems of the crop plants or fruit trees are called lenticels. The process of water evaporation through these lenticels is known as lenticular transpiration. The water lost by this process is extremely less ie., about 0.1 percent.

Lethal

So detrimental as to cause death.

Level of free convection (LFC)

The level at which a parcel of saturated air becomes warmer than the surrounding air and begins to rise freely, in a conditionally unstable atmosphere.

Levelling

The tillage operation in which the soil is moved to establish a desired soil elevation slope.

Ley farming

A cultivation that includes rotation of annual arable crops and artificial pastures occupying field for two years or longer.

Lidar

Light amplification by stimulated emission of radiation is abbreviated as 'LASER'. If this pulsed laser radiation is back scattered by clouds, dust and others in the atmosphere, it is termed as 'Lidar'.

Liebig's law of minimum

The growth and reproduction of a crop plant is dependent on many inputs, including essential nutrients and minor elements that are available in minimum quantity.

Life belt

The vertical sub-division of crop plant life, which is largely determined by latitudinal influences.

Life saving irrigation

The irrigation given to a dry land/rainfed crop to save it from total failure due to prolonged dry spells or severe soil moisture stress, the source of irrigation generally being harvested water.

Lift

The very commonly used term in all the synoptic measurements in meteorology. The force applied in perpendicular direction against a balloon from which the readings are observed is termed as 'lift'. So, to support the weight of the balloon in air, this component is effectively useful.

Lifted index (LI)
A measure of atmospheric instability that is obtained by computing the temperature that the air near the ground would have if it were lifted to a higher level and comparing it to the actual temperature at that altitude. Positive values indicate more stable air and negative values indicate instability.

Lifting condensation level
The level at which an air mass becomes saturated after being lifted adiabatically. Here, it is to be noted that the conservation of vapour takes place.

Light air
The speed of the wind is between 1 and 3 miles per hour and lies in between calm and light breezes. All these terms are used only to describe the wind force based on the intensity.

Light breeze
Wind with a speed of six to eight nautical miles per hour or two on Beufort scale wind force.

Light compensation point
The light intensity at which photo-synthesis equals respiration.

Light soil
A soil that is very easy to work with implements or easy to cultivate as it lacks cohesiveness, such as sandy loam soils.

Light waves
The part of the electromagnetic spectrum of solar radiation that contains visible light.

Light
This is the part of electromagnetic spectrum in the wave lengths between 0.4 and 0.7 microns. This 'light' segment which exists between the ultraviolet and infrared parts of the spectrum is sensitive to the human eye and much of the terminology concerning solar radiation is based on the relationship between wavelength and colour. Different wavelengths within the visible part of the spectrum have particular colours such as violet, blue, green, yellow, orange and red.

Lightning

Luminous manifestation accompanying a sudden electric discharge in the form of a magnificent curvy flash or spark which exists in between two clouds, inside a cloud, from a high structure on the ground or even between the cloud and the Earth.

Line echo wave pattern (LEWP)

A wave or "S" shaped bulge in a line of thunderstorms associated with severe weather.

Linear acceleration

Acceleration is the rate of change of velocity. If the acceleration is in the direction of flow of a fluid then it is said to be a linear acceleration. Here, it is to be remembered that the 'fluid' means both gases and liquids.

Linear perspective

The depth cue of parallel straight lines seeming to converge at a distant vanishing point.

Liquid fertilizer distributor

A machine for applying mineral fertilizer in liquid form which either spreads the fertilizer on the surface of the ground or injects it into the soil.

Liquid fertilizers

Commercial fertilizers in liquid form. Such fertilizers are chiefly anhydrous ammonia, aqueous solution of nitrogen, and some mixed fertilizers. Liquid fertilizers are applied to the soil through irrigation water, as starter solutions, and with the help of special equipment.

Liquid manure

Liquid resulting from animal urine and litter juices or from a dung heap.

Liquid thermometer

A thermometer in which liquids like ethyl alcohol, petroleum, etc., are used and temperature is measured with the rates of expansion and relevant change in temperature.

Lithometeor

The dry particles that hang suspended in the atmosphere, such as dust, smoke, sand, haze, etc.

Lithosphere

The Earth's crust, consisting of the surface soil lying upon the hard rock which is several kilometers thickness. The crops are raised on fields, which are the components of this part of the globe.

Littoral coast

Generally, a coastal climate is warmer, has more relative humidity, uncomfortable summer days, sultry, etc., in the tropics, semi-arid tropics and in similar parts of the world. If such a coastal climate on the continents is subjected to change by water, then the situation is said to be 'littoral coast'.

Livingston atmometer

See 'Livingston sphere'.

Livingston sphere

A devise used to measure the evaporation and is also known as Livingston atmometer.

Loam

Soil material that contains 7-27 per cent clay, 28-50 per cent silt and less than 52 per cent sand.

Local apparent time(LAT)

The interval between two successive transits of the Sun across the meridian is called the true solar day and the time based on the length of this day is called 'apparent solar time'. Local apparent time is the apparent solar time of any particular place such that the Sun passes across the geographical meridian at noon.

Local control

Making use of greater homogeneity of adjacent areas in a field, the entire experimental units are divided into compact blocks which is known as local control. Local control will reduce the experimental error thereby increasing the precision of the experiment.

Local mean time (LMT)

This is the local time based on the transit of the mean Sun. To calculate LMT from IST, it is essential to know the longitude of the station.

Local winds

According to the change in pressure and convection properties, alongwith some other climatic parameters the movement of air takes place in the atmosphere. If such air is horizontal in relation to overlying ground then it is said to be wind. Under a given situation over a small area, preferably in an area of 2 to 3 villages, if the winds differ from general circulation of comparatively larger area like 2 to 3 districts, then the winds over the villages can be called as local winds. This is more a relative term and differ largely from the view point of one scientist to another. The description given above may positively hold good for agrometeorologists in general and forest micrometeorologists in particular.

Localized placement of fertilizers

This refers to the application of fertilizer in to the soil close to the seed or plant. This method is specifically adopted with phosphatic fertilizers for three reasons, namely, restricted contact of fertilizers with soil lessens fixation phosphate, necessary plant food is placed within easy reach of plant roots, and application of fertilizers in a band along the rows does not readily furnish nutrients to weeds growing between the rows.

Lodging (crop lodging)

A state of standing crop in the field being fallen over before harvest by heavy rain and/or wind either as a result of excessive soil fertility, particularly levels of available nitrogen or the presence of fungal disease on the stem.

Logarithms

If a $P = N_1$, p is called the logarithm of N to the base a, written as p = log a N. If a and N are positive and a is not equal to 1, there is only one real value for p. The following rules are important.

(a) log a MN = log a M + log a N

(b) log a M/N = log a M – log a N

(c) log a M^n = n log a M

In usage two bases are used, the Briggsian system, uses a = 10, and the Naparian system uses the natural base a = e = 2.7182.

Long day plant

Plant in which flowering can be induced or enhanced by long days. The condition is that the day should have more than twelve hours of day light, e.g., barley, potato, etc.

Plants in which flowering is initiated or hastened under day length above critical value e.g., wheat, barley, oats etc. Long day plants otherwise called short night plants.

Long range forecast

The weather forecast valid for the next three to four months of the season. The forecast is made using global numerical models, the use of which varies from planning to re-planning the seasonal operations of agriculture to a weather based agricultural management system. The other industrial and socioeconomic advantages are also numerous. A type of weather forecast for agriculture the validity of which is generally to be extended from a month to a season.

Long wave radiation

The terrestrial radiation in the wave-lengths between 4 and 100 microns and in some specific instances upto 120 microns of electromagnetic radiation is also considered as long wave radiation. These wavelengths are emitted by the Earth and its atmosphere. This radiation is emitted after the Earth is heated by solar radiation and become a source of radiation itself. The important aspect of this radiation is that the absorption and emission of radiation in atmospheric gases occurs in discrete wavelength bands but not uniformly across the spectrum. This is invisible radiation.

Long wave trough

A wave in the prevailing westerly flow aloft characterized by a large length and amplitude, which moves slowly and is persistently.

Logarithmic velocity profile

A logarithmic scale of measurement is one in which an increase of one unit represents a tenfold increase in the quantity measured. The rate of motion in a given direction is velocity. This term is normally used in the studies of surface layer fluxes and gradients under different situations. In most of the micrometeorological studies of crop plants the speed of the wind is related to logarithm of height above the canopies. It is believed that the wind speed increases with the logarithm of height over most of the crops. The wind velocity profile studies enable the micrometeorologists to estimate friction velocity, eddy diffusivity, surface stress, etc., if wind speed at two to three heights over the crop is known alongwith surface roughness. In one line, this term can be stated as the profile in the surface boundary layer. In other instances, this is also abbreviated as 'log wind profile'. So, whatever may be the name given, one can see the linear profile of wind speed against log height in these profiles.

Longitude

In order to determine the exact position of a point on the surface of the earth both latitude and longitude are required. A plane passing through both the poles cuts the Earth in a great circle known as meridian. A meridian can be drawn through any point on the surface of the Earth. The longitude of a point on the surface is the angle between its meridian and a standard meridian passing through a given point at Greenwich in England. The Greenwich meridian is called the 'Prime Meridian', or '$^{\circ}$' longitude. As one moves eastward or westward from Greenwich meridian the longitude values increase until 180° is reached. For every 1° difference in longitude east of Greenwich 4 minutes is to be added to the time of Greenwich, India is based 82.5° E (Allahabad).

Longitudinal axis

An axis in the longitudinal direction of the figure or body considered, and generally passing through the center of gravity or the center of figure.

Lorentz contraction

To correlate between space and time equations in two frames of references at realistic velocities, Hendrik Lorentz (1853-1928) developed a set of equations. Likewise, Lorentz contraction is the relativistic contraction, at almost the speeds of light, of distances parallel to the direction of motion.

Low clouds

The clouds which are present in the atmosphere upto a height of about two thousand metres from the surface of the Earth. On a few occasions the minimum height is as low as zero metres. Perhaps, the fog is considered as a cloud itself and one can feel the conditions of a cloud while moving in a fog. It is to be remembered that there are different forms of fog.

Low latitudes

The latitude belt between 30 and 0 degrees north and south of the equator.

Low Level Jet (LLJ)

Strong winds that are concentrated in relatively narrow bands in the lower part of the atmosphere.

Low pressure area

See 'Cyclone'. 'Tropical storm' and 'Depression'.

Low pressure system

An area of a relative pressure minimum that has converging winds and rotates in the same direction as the earth. Also known as cyclone.

Low Sun

Like 'high Sun' this term is used to indicate the season when the Sun is at its minimum noon altitude.

Low

In a system of winds, an area of low pressure. Also known as depression or cyclone.

Lucimeter

An instrument used to measure the global solar radiation.

Lumen

The S.I. unit of luminous flux. This is equal to the amount of flux intercepted by one square foot of surface held perpendicular at a distance of one foot from a standard candle. Also see 'Luminous flux'.

Luminescence

The emission of light that is not caused by high temperature.

Luminous body

A body which emits light.

Luminous flux

The amount of light passing through a unit area in one second. The S.I. unit is lumen.

Luminous intensity

The luminous flux per unit solid angle of the direction in question.

Lunar Eclipse

An eclipse of the moon which occurs when the earth is in a direct line between the sun and the moon.

Lunar periodicity

The correlation of activities of certain organisms with periods of moon. This happens in certain crop plants, but there is not authentic scientific proof as on date.

Lux

The derived S.I. unit of illumination. One lumen per metre per metre. In other words, 'Metre candle'.

Lysimeter

An instrument used to measure the loss of water from the soil. In this instrument the precipitation, condensation and irrigation are considered as gains and the evapotranspiration is considered as loss.

M

M

number of weeks in the crop growth period/season.

M.K.S. system

A system of measurement in which the fundamental units are metre, kilogram and second for length, mass and time respectively.

Mackerel sky

The name given to cirrocumulus clouds with small vertical extent and composed of ice crystals. The rippled effect gives the appearance of fish scales.

Macroburst

A large downburst with an outflow diameter of more than 4 kilometers with damaging winds.

Macroclimate

Climate of a continent or over the complete globe.

Macroenvironment

The total of abiotic and biotic environments which surround the crop field and its microenvironment.

Macrometeorology

A branch of meteorology concerned with the study of large scale meteorological phenomena over a continent or the entire globe.

Macronutrient

A chemical element necessary in relatively large amounts (usually 500 parts per million in the plant) for the growth of plants. These elements consists of C, H, O, Ca, Mg, K, P, S, and N.

Macroscale

The meteorological scale covering an area ranging from the size of a continent to the entire globe.

Magnesium sulphate

Usually the product known as kieserite ($Mg\ SO_4\ H_2O$). The highly hydrated form Epsom slats ($MgSO_4\ 7H_2O$) is suitable for foliar application.

Magnetic poles

Either of the two points on the earth's surface where the magnetic meridians converge.

Magnetosphere

The space surrounding any celestial body and also the Earth which possesses magnetic field with them.

Mamma

Under unstable conditions of the atmosphere, the bulbous protuberances formed on cumulonimbus clouds.

Mammatocumulus

A portion of a cumulonimbus cloud that appears as a pouch or udder on the under surface of the cloud, which indicate the occurrence of a storm.

Manometer

A gauge for measuring the pressure of gases, vapours and fluids.

Manure

The excreta of animals–dung and urine, with straw or other materials used as the absorbent. The decomposed manure is called farmyard manure. The average composition of well-decomposed farmyard manure is 0.5 per cent nitrogen, 0.3 per cent P_2O_5 and 0.5 per cent K_2O.

Mare's tail

The cirrus clouds composed of ice crystals that appear as veil patches or strands, resembling the tail of a horse.

Marginal visual flight rules (MVFR)

Refers to the general weather conditions pilots can expect at the surface with 3-5 miles visibility.
Related terms : IFR , VFR

Marine climate

Almost similar and on many occasions used as a synonym to maritime climate. The climate is basically influenced by adjacent seas and oceans. See ; 'Maritime climate'.

Maritime air

An air mass containing high level of moisture accumulated over a period of several days because of its existence over seas or oceans. Even though the chances of uniformity of moisture content in the total air mass is less, atleast, at the lower levels it exists to a considerably extent.

Maritime climate

The climate experienced in areas and regions near the oceans and seas. It is characterized by hot dry summers and mild wet winters. The diurnal variations of temperature is small and air is with high relative humidity.

Maritime

Air (masses) originating over the seas and oceans.

Markov chain
The probability of occurrence of dry and wet spells are developed by many scientists in meteorology. This model is the best as far as its applicability is concerned, in all the agrometeorological problems.

Mass density
The mass per unit volume of any matter or material. Also see 'Mass'.

Mass
The quantity of matter or material in a body, a measure of its gravitational influence and also its inertia.

Mathematical model
A mathematical model is an equation or set of equations which represents the behaviour of a system. The equations which were developed based on the mathematics can be used in these models.

Matter
The material of which substances are made. This is a specialized form of energy (actually energy acts on matter) that has the attributes of mass and extension in space and time.

Mature soil
The soil that is in good adjustment with environmental conditions in many regions with well developed horizons.

Maturity or mature grain stage
When grain colour in the panicle begins to change from green to yellow and 90-100 per cent grains are fully developed, hard and free from green tint.

The stage which a plant is capable of reproducing itself by seed.

Maunder minimum
The British solar scientist 'Alter Maunder' observed very few sunspots between 1645 and 1715 GMT. These are the minimum that could be seen and this is named after this great scientist.

Maximum available water
The quantity of water represented as the difference between field capacity and wilting coefficient. This is the moisture that can be readily extracted by the plant and used for normal growth.

Maximum residue limit (MRL)

Means the maximum concentration of a residue that is legally permittted or recognized as acceptable in or on a food, agricultural commodity or animal feedstuff.

Maximum temperature

As per the IMD guidelines, the highest temperature recorded during a day, week, month, season, etc.

Maximum thermometer

A thermometer used to measure the highest air temperature attained in a day or a given interval of time.

Maximum tillering stage

The stage of the crop at which it has produced maximum potential number of tillers. It is the main yield attributing character.

Maximum

The greatest value attained by a weather element. (temperature, pressure, wind speed, humidity etc).

Maxmium cropping

The highest possible production per unit area per unit time without considering cost of production or net return.

Meadow

Area covered with grasses and/or succulant forage legumes grown primarily for hay.

Mean daily maximum temperature

As per the IMD guidelines, the mean value of the recorded daily maximum temperature of any calendar month in a year, or for a specified month required.

Mean daily temperature

As per the IMD guidelines, the mean of the temperatures obtained for 24 continuous recordings at hourly intervals in a day.

Mean free path

In the kinetic theory of gases, it is assumed that between two collisions, the molecules travel a certain distance in a straight line. The mean free path to the mean distance travelled by a gas molecule between two successive collisions.

Mean monthly maximum temperature

The mean value obtained arithmetically and recorded for maximum temperatures observed in a calendar month or for a specified month required.

Mean sea level

The average height of the sea surface water level.

The mean place at which the tide oscillates: the average height of the sea for all stages of the tide. At any particular place, it is derived by averaging the hourly tide heights over a 19-year period.

Mean square

The variance so called because it is the average of the squared deviations from the arithmetic mean of a series of values.

Mean temperature

The average of temperature readings taken over specified amount of time. (The average of the maximum and minimum temperatures during a day).

Mean

The mean is the arithmetic average and is obtained when the sum of the values of individuals in the data is divided by the number of individuals in the data.

Measured ceiling

A ceiling classification applied when the ceiling value has been determined by an instrument, in other than natural landmarks.

Mechanical analysis

The determination of the relative distribution of the size group of ultimate soil particles. The important steps in mechanical analysis are:

1. separation of all particles from each other so as to cause complete dispersion into ultimate particle
2. measuring the amounts of each size group in the sample.

Mechanical convection

When the turbulence is caused in the atmosphere by vertical wind shear, the vertical exchange of air parcel takes place. The process is termed as mechanical convection and as a result of this process the storms occur.

Mechanical equivalent of heat

This is a constant in heat and work relationship and is the amount of work required to produce a unit quantity of heat. This is also called as Joule's equivalent. One calorie at 15 degree centigrade is equal to 4.185×10^7 ergs. One B.T.U. is equal to 778 ft.lb. So, mechanical equivalent of heat has the values of 4.185×10^7 ergs per calorie and 778 ft.lb.

Mechanical farming

It is the farming in which machine drawn implements are used for the reduction of labour requirement or elimination of manual work, timeliness of operations and improved quality of husbandry resulting in higher output and better quality of produce for increased profit.

Mechanical turbulence

Forced turbulence is also called as mechanical turbulence. See 'Forced turbulence'.

Mechanics

The branch of physical sciences dealing with the effect of forces on bodies.

Mechanistic model

These are based on known principles and attempts to give a description of the system with understanding . Whatsoever may be the basic principle involved that can be used in these models.

Mediterranean climate

A climate type, of which, the Mediterranean basin is the archetype, and is characterized by hot dry summers and mild wet winters.

Medium

A substance which is subjected to action and through which waves travel or pass through.

Medium range forecast / Extended forecast

The weather is forecast both by synoptic and statistical approaches, the purpose of which is to enable the planners to make a complete programme for taking control measures and for taking such steps which may suppress the spread of diseases. The medium range forecast is the weather forecast for the next 3 to 10 day period, from the day of issue.

A type of weather forecast for agriculture the validity of which is generally considered to be extended from 3 to 10 days. Even though there is no standard limit the validity upto 5 days is taken as the agreed limit for high level of accuracy. Medium range forecast is also known as 'Extended forecast'.

Mega thermal climate

This climate is characterized by warmness and uniformly high temperature and is winterless. Heavy rain occurs atleast in one month of a year and in some regions the rainy season may occur twice in a year.

Mega watt

A mega watt is one million watt. Watt is named after James Watt (1736-1819). Also see 'Watt'.

Melting level

The level at which ice crystals and snow flakes melt in their descent through the atmosphere.

Melting point

The temperature at which a solid substance changes to a liquid phase, and this is a function of pressure. At melting point, the temperature is constant and the solid and liquid phases, of a substance are in equilibrium. The melting points are quoted for standard atmospheric pressure, i.e., 760 mm of mercury.

Melting

The change of ice to liquid water.

Mercurial barometer

The barometer in which pressure is determined by balancing air pressure against the weight of a column of mercury in an elevated glass tube.

Meridian (Terrestrial).

Meridian of longitude. An imaginary great circle drawn on the Earth through both the poles.

Meridional flow

Atmospheric circulation in which the north and south (meridional) component of motion is pronounced.

Meridional

Along a line of meridian of longitude.

Mesoclimate

The climate of a region or location of small size like a valley, a forest, etc. In mesoclimatic studies, the exchange of heat, moisture and momentum between these locations and atmosphere are studied. The mesoclimate is intermediate in scale between the macroclimate and microclimate.

Mesoclimatology

There are many definitions. Of them, the following are largely accepted.

- A branch of climatology concerned with the study of mesoclimates of certain specific locations, their effects on natural vegetation, etc.
- A branch of climatology that studies the mesoclimates.

Mesocyclone

An area of rotation of storm size found on the southwest part of a super cell.

Mesohigh

A small, concentrated area of high pressure created by the cold out flow and rain-cooled air from thunderstorms.

Mesolow

A small scale low pressure center, of the size of an individual thunderstorm to many tens of miles.

Mesometeorology

A branch of meteorology concerned with the study of the medium scale meteorological phenomena like sea and land breeze, a tornado, a convective storm, etc. The area of study is of the order of a hundred or a few hundreds of square kilometers.

Mesopause

The layer of the atmosphere, lying between 50 and 80 kms above the Earth's surface, separating mesosphere from thermosphere. In this layer, the temperature decreases with height.

Mesoscale convective complex (MCC)

A large mesoscale convective system (MCS) of the size of the few thousand square kilometres and lasts at least 6 hours.

Mesoscale convective system (MCS)

A large organized convective weather system comprised of a number of individual thunderstorms.

Mesoscale

The atmospheric phenomena occurring upto 100 km above the Earth over a valley, etc. Some of these are cold and warm fronts, sea and land breezes, storms, tornadoes, etc.

Mesosphere

Of the different layers of the atmosphere, this is the layer above the stratosphere, lying from about 18 to 80 kilometres. In this region of the upper atmosphere, the temperature decreases rapidly with height.

Mesotherm

The vegetation that requires moderate warmth and moderate moisture. For instance, a maize crop canopy is a classical example for this category. This type of vegetation is found in regions lying between 22 and 45° N and S of the equator.

Mesothermal climate

Climate characterized by moderate heat and moisture. Mild winters are archetype in this climate and winter and summer seasons are clearly found.

Metar

Acronym for *M*eteorological *A*erodrome *R*eport.

Meteorograph

An apparatus for automatically recording two or more meteorological elements simultaneously. This apparatus may be kept either in a cropped field or in the observatory.

Meteorological drought

If annual rainfall is significantly short of certain level (75 percent) of climatologically normal rainfall over a wide area, then the situation is called by this term. It may be noted that in every state each region receives certain amount of normal rainfall. This is the basis for planning the cropping pattern of that region or area.

Meteorological observatory

An observatory in which meteorological instruments are installed and correct readings are observed, recorded and maintained in specified books, as per the IMD guidelines.

Meteorology

The commonly accepted definitions :

- The science concerned with the study of the characteristics and behaviour of the atmosphere.
- The science of the atmosphere.

Meteorotropisms

The Greek term 'meteor' refers to atmospheric disturbances and 'tropic' to turnings or changes. Hence, this term refers to a turn of events related to atmospheric disturbances.

Metre

The unit of length in metric system and equal to one hundred centimetres. The symbol is 'm'.

Microbarograph

An aneroid barograph used to observe and record atmospheric pressure changes of very small magnitude.

Microburst

A severe wind blasting down locally (4 Km) from a thunderstorm.

Microclimate

In meteorology, the climate experienced in a valley of small area, when compared to that experienced over the entire mountain. However, in agricultural meteorology the microclimate has different meanings. One version is that the climate experienced in the air space above the Earth to a height upto which the characteristics are homogenous as compared to the immediate other surface above it. The second version is a continuation of the above, plus, the depth in the soil upto which the root zone of a crop exists and also just above the crop canopy, (ie., the ecosphere of the crop is often taken into consideration).

Microclimatology

As such there is no standard and widely accepted definition. However, narrations are available to give a clear meaning to this term, when compared with micrometeorology. Microclimatology deals with diurnal and seasonal variations as well as long term trends in meteorological parameters in general in the near surface layer of the atmosphere, and the long term averages of shortwave radiation, wind velocity, temperature, humidity, etc., in particular. It is to be mentioned very clearly that both the branches deal with similar atmospheric processes occurring near the surface of the Earth.

Microenvironment
The immediate surroundings of a crop and its canopy.

Micrometeorology
There are various definitions

- A branch of applied meteorology concerned with the detailed examination on a micro scale basis of the physical and meteorological processes that are taking place within the ecosphere.

- A branch of applied meteorology concerned with the study of small scale meteorological conditions which involve refined measurements close to the surface of the Earth over relatively short periods of time.

Micro-micron
A metric measure, one millionth part of the micron.

Micro-millimetre
A metric measure, one millionth part of a millimeter.

Micron
A unit of length used to designating the wavelength of light. This is equal to one millionth of a metre.

Micronutrient (minor element or trace element)
A chemical element necessary only in extremely small amounts (usually less than 50 parts per million in the plant) for the growth of plants. These elements include B, Cl, Cu, Fe, Mn, Mo and Zn.

Microphysics of clouds
The study of physical processes on the micro scale of individual cloud, such as precipitation droplet, size, etc.

Microplot
One of many similar small plots used in determining accurately the comparative performance of varieties in the field.

Micropluvimeter

The rain gauge used to register very light precipitation.

Micropores

Soil characterising pores size having diameter ranging from 3 to 30 microns which plays an important role in water and nutrient retension with slow capillary flow.

Microscale

The smallest scale of meteorological phenomena that range in size from a few centimetres to a few kilometres.

Microtherm

A plant that can grow in cool and short summers with low means values of annual temperature. The monthly mean temperature for the warmest month is atleast 10° and less than $22\ ^{o}C$.

Microthermal climate

Basically, cold climate that controls the region. Cool and short summers and winters are colder than other seasons. Snow may occur once in a while and it rains in the warm season.

Microwave

The portion of the electromagnetic spectrum with wavelengths ranging from 0.3 to 100 centimetres.

Mid-day depression or mid-day water deficits

During periods of high radiation and low humidity even those plants growing in soils nearing field capacity may be subject to severe water stress. The phenomenon is known as mid-day depression or water deficits.

Middle couds

A term used to signify clouds with a hight ranging from 2.5 to 7 Km.

Middle latitudes

The latitude belt between 35 and 65 degrees North and South, of the equator.

Minimum

The least value attained by a weather element (temperature, pressure, wind speed,humidity etc).

Mid-season correction

It is a contingent managemet practice to overcome unexpected and or unfavourable weather conditions e.g., thinning or population adjustment according to the availability of moisture in the soil during the dry spell.

Mie scattering

This is a phenomenon in which scattering of radiation by particles with circumference of more than thirty times of the wave length of incident radiation. White light is scattered, because, in such cases the scattering is independent of the wavelength.

Milankovitch periodicities

The time bound regular changes in the orbital phenomena of the Earth which control the onset of ice ages.

Milk grain stage

The stage where the contents of the caryopsis (the starch portion of the grain) are first watery but later turn into milky white liquid which can be squeezed out. It is critical stage for water, weed conditions.

Milli watt hour (mwh)

The unit for measuring radiant energy and is equal to one thousandth of a watt hour.

Milli watt

The derived SI unit for measuring power or intensity of solar radiation. A watt is equal to one Joule per second and a mili watt is one thousandth of a watt.

Millibar

The meteorlogical unit of pressure. This expresses directly, the force exerted by the atmosphere ; equivalent to one thousand dynes per centimetre per centimetre or one hundred pascals.

Milling recovery

Weight of commercial rice recovered from the original rough rice (Paddy) usually expressed in per cent. Generally it is about 48%.

Minimum temperature

As per the IMD guidelines, the lowest temperature recorded during a day, week, month, a season, etc.

Minimum thermometer

A thermometer used to measure the minimum air temperature attained in a day or a given interval of time.

Minimum tillage

The soil manipulation necessary to meet the minimum tillage requirements for crop production. In this method weeds are controlled by herbicides only.

Mist

An air mass close to the Earth's surface which has been cooled to below its dew point. The result is condensation into minute water droplets. Visibility in mist is under 2000 m (6000 ft). Thus, the difference between fog and mist is basically the degree of visibility.

Mixed farming systems

Farming systems with integrated crops, livestock, and other possible household enterprises.

Mixed farming

Farming systems which involves the raising of crops and rearing of animals and/or poultry. Mixed farming is based upon the principle that land should support animal and animal should support land.

Mixed layer

The upper portion of the boundary layer in which air is thoroughly mixed by convection.

Mixed phase

A phase in which the water exists simultaneously in its three states viz., solid, liquid and vapour.

Mixed precipitation

A combinations of liquid and solid precipitation.

Mixing ratio

The ratio of the mass of water vapour to the mass of dry air occupying the same volume. This term is more strictly 'water vapour mixing ratio' and for all purposes applied in a system of moist air.

Moderate breeze

Air in motion with speed between thirteen and sixteen nautical miles per hour or four on Beufort scale wind force.

Moist adiabat

The line on a Skew T-Log P chart depicts the change in temperature of saturated air as it rises and undergoes cooling due to adiabatic expansion,

Moisture air

Air of high relative humidity in which both dry air and water vapour are present.

Moisture availability index

It is the ratio of assured rainfall (weekly or monthly) at 50% probability level to potentital evapotranspiration of the corresponding period.

Moisture content (soil)

The quantity of water present in soil usually expressed in percentage by weight, on oven dry basis. It can be calculated by volumetric and gravimetric method.

Moisture deficit index (MDI)

This is the measure to estimate dryness of a region and is expressed by the formula $\dfrac{P - PET}{PET} \times 100$

where

P = Precipitation

PET = Potential evapotranspiration

Moisture equivalent

The moisture retained in air-dried, screened sample of soil which has been wetted and drained in a standard manner and centrifuged for 30 minutes in a centrifugal field equal to 1,000 times gravity. Moisture equivalent is expressed as moisture percentage on a dry-weight basis and approximate field capacity for many medium and fine textured soils.

Moisture index (MI)

The potential evapotranspiration is to be deducted from precipitation and simultaneously divided by PET. Then this value has to be multiplied by one hundred, to arrive at the MI.

Moisture regimes

The different levels of the available soil moisture (ASM as%) a testing factor or unit inputs used in conducting irrigation experiments, such as ASM at 100, 80, 60, 40 per cent etc.

Moisture release curve or characteristic curve

It is the functional relationship between soil moisture tension and soil moisture content at a range from field capacity to wilting point.

Moisture retentive soil

A soil that retains a high percentage of water directly depending upon clay colloid content, clay soils are more moisture retentive than sandy soils.

Moisture stress

The tension at which water is held by the soil.

Moisture tension

The equivalent negative pressure to which water must be subject in order to be in hydraulic equilibrium through a porous permeable wall or membrane, with the water in the soil.

Moisture

The water vapor content in the atmosphere.

Molasses

By-product of sugar industry, viscous liquid left behind after sucrose is extracted, separated by centrifuging sugar in sugar factories, used in preparation of methylated spirit, power alchohol, or alcoholic beverages, yeast, cattle feed etc.

Mole

The basic SI unit of amount of substance. A quantity of any substance such that its mass in grams is expressed by the same number as the average atomic or molecular mass. Therefore, a mole of any pure material contains Avogadro's number of atoms or molecules, i.e., 6.023×10^{23} molecules.

Molecule

The smallest portion of a substance which retains not only the properties of the substance but also the capability to exist independently.

Moment of a force

A measure of the tendency of a force to rotate the body, i.e., its turning effect. This is measured by the force times its perpendicular distance from the line of action of the force to the axis of rotation.

Momentum

This is the property of a particle and is the product of mass and velocity. A vector quantity.

Monetary advantage index (MAI)

While assessing yield advantages of different intercropping systems, an index was formed by taking into consideration the relative money value of the produce.

Monin

This is a scaling length for surface layer turbulence. Under stable conditions of the atmosphere this is positive, negative under unstable and infinite for neutral conditions.

Monochromatic emissivity

The ratio of the monochromatic emissive power of the body to the monochromatic emissive power of a black body at the same wavelength and temperature.

Monochromatic Light of one colour and consisting of vibrations of the same single frequency.

Monocrop (mono culture)

A cropping system in which the same crop is grown year after year in the same field.

Monsoon climate

Any climate similar in nature to that experienced in the Indian sub-continent during monsoon.

Monsoon effect

A tropical forest of deciduous trees in regions where seasons of heavy rainfall alternate with long droughts.

A forest type found in areas having a well marked rainy season and characterized in some regions by tree species that loose their leaves for a least of the dry season.

Monsoon season

The period during which the continental monsoonal winds blow.

Monsoon winds

These are the trade winds associated with monsoon, that change their course and direction periodically with the seasons. The Indian sub-continent has a well developed regular monsoon system. Also see 'Monsoon'.

Monsoon

The name is derived from the Arabic word "Mausin' meaning a season. Monsoon is a large scale flow of seasonal winds of persistent direction either from ocean on to a heated continental land mass (in summer) or from the land mass to ocean (in winter), particularly on the Indian sub-continent. In these events a pronounced change in the direction of seasonal (reversal) trade winds occur. The phenomena, obviously, take place as a result of temperature difference between land and water. The monsoon period is very wet and frequently stormy weather is experienced. The name is used sometimes for similar but weaker behaviour in higher latitudes.

Monthly maximum temperature

As per the IMD guidelines, in any given year, the highest temperature recorded in a calendar month.

Monthly minimum temperature

As per the IMD guidelines, in any given year, the lowest temperature recorded in a calendar month.

Molal concentration

The number of gram molecules of the solute present in 1000 grams of solvent, usually water.

Molal solution

A solution in which one gram molecular weight of the substance is dissolved in the solvent and made upto one litre.

Motion

Movement from one place to another. Motion equations are those that apply to bodies moving with uniform acceleration.

Mound culture

Soil is separated into mounds varying in height (30-50 cm) and shape to fit local needs. Distance between mounds are determined by soil depth, amount of soil needed to construct mounds of a given height and type of crop to be grown.

Mountain breeze

A breeze blowing downslope of a mountain.

Mountain climate
The climate experienced on the mountain which changes with altitude. There may be certain other factors. But, all those occur mainly due to change in the altitude.

Mountain parallax
A relative motion of near and far objects perceived when either the object or the observer is moving.

Mountain wave
A wave in the atmosphere caused by the obstruction of mountain.

Mountain wind
The descending wind flow from a mountain which may be either cold or slightly warm depending on the pressure, temperature differences, etc.

Muck soil
Fairly well decomposed organic soil material relatively high in mineral content, dark in colour, and accumulated under conditions of imperfect drainage.

Mud slide
Fast moving soil, rocks and water that flow down mountain slopes during a heavy downpour of rain.

Muggy
The warm and excessively humid weather.

Mulch farming
A system of farming in which the organic residues are not ploughed in to the ground but are left on the surface.

Mulch tillage
Preparation of soil in such a way that plant residues or other mulching materials are specially left on or near the surface.

Mulch
Any material such as straw, plant residues, leaves, loose soil or plastic film placed on the soil surface to reduce evaporation, erosion or to protect plant roots from extremely low or high temperature.

Multicel storm

A thunderstorm made up of two or more single-cell storms.

Multiple cropping index

Measures the sum of area planted to different crops and harvested in single year divided by the total cultivated area times 100.

Multiple cropping

Growing two or more crops consecutively on the same field in the same year. In essence represents a philosophy of maximizing crop production per unit area of land within a calender year with minimum deterioration of soil.

Multiple vortex tornado

A tornado which has two or more condensation debris clouds.

Multi-storyied/multi-tier cropping

Is a system of growing together, crops of different heights at the same time on the same piece of land and thus using land, water and space most efficiently and economically i.e. coconut + pepper + pineapple + grass.

Mungari

In Andhra Pradesh the early crop season extending from June to October.

Munsell colour system

A classification of colour in terms of three attributes, hue, clue, and chroma. By this system, any colour can be specified by numbers and letters.

Muriate of potash

Commercial potassium chloride of 90 to 99 per cent grade. It contains nearly 60 per cent K_2O.

N

Nadir
The point on any given obsever's celestial sphere diametrically opposite of one's zenith.

Nanometre
The unit of length specifically applied to wave-lengths of radition. This is equal to ten Angstrom units or 10^{-9} metre.

National agricultural research project (NARP)
Is a project launched by the ICAR to strengthen the regional research capabilities of the agricultural universities in India, with an objective of conducting location-specific and production oriented research in all the agroclimatic zones identified in the area of its jurisdiction.

Native pasture
A pasture covered with native plants or naturalized exotic plants.

Natural convection
See 'Free convection'.

Natural numbers
The numbers 1, 2, 3, 4, etc., through which the process of counting certain sets of objects is done and the totalities of these numbers.

Nautical mile

A unit of length used in marine navigation which is equal to a minute of arc of a great circle on a sphere. One international nautical mile is equivalent to 1,852 meters.

Nautical twilight

The time after civil twilight.

Neap tide

A tide of decreased range, which occurs about every two weeks when the moon is at one quarter or three-quarters full. Related term: Spring tide

Near gale

Air in horizontal motion with a speed between twenty seven and thirty one nautical miles per hour or seven on Beufort scale wind force.

Near infrared radiation

The wavelengths of radiation nearest in magnitude to visible light. In the solar spectrum the spectral range is between 0.7 and 1.5 microns.

Negative phototropism

The phenomena in which the shoots of a crop plant will not turn towards light even under increased intensities.

Negative specific heat of saturated vapour

If saturation has to be maintained at higher temperature in an air parcel containing vapour, the heat has to be withdrawn. So, for a saturated vapour parcel this component is always negative.

Negative vorticity advection

The advection of lower values of vorticity into an area.

Negatively charged body

A particle or molecule with excess of electrons.

Nephanalysis

An analysis of cloud to determine its cover over the sky and also its type including height.

Nephelococcygia

A term that applies when people find familiar objects with the shape of a cloud.

Nephelometer

An instrument used to measure the scattering function of particle present in a medium.

Nephelometry

A study to determine the scattering properties of a relatively small sample of air.

Nephology

The study of clouds including the microphysics of clouds and their characteristics

Nephometer

An instrument used to measure the amount of cloudiness of the sky. Usually, this is measured in oktas. Also see 'Clear sky' , 'Overcast sky', etc.

Nephoscope

An instrument to detect and observe the direction of the cloud motion

Net area sown

Class of land under revenue classification represents actual extent of land sown to crops during a given year, when cropped more than once in any year, its area is added to the net area sown to arrival at the total area sown during the year.

Net assimilation rate

The increase in dry weight of a plant and is expressed in grams per metre per day. This is also known as 'Unit leaf rate'.

Net assimilation

The difference between the assimilation and respiration.

Net energy (NE)

Difference between metabolizable energy and heat increment; includes the amount of energy used either for maintenance only, for maintenance plus production.

Net irrigation requirement (NIR)

Is the depth of irrigation water exclusive of precipitation, carry-over soil water or ground water contribution or other gains in the soil moisture, that is required for plant growth; the amount of irrigation water required to bring the soil moisture level in the effective root zone to the field capacity.

Net long-wave radiation

The difference between the total incoming long wave radiation and the total out going long wave radiation. But, while measuring with any instrument of even greater accuracy, the total incoming short wave plus long wave radiation and the total out going short wave plus long wave radiation should be taken into consideration. The value of the net long wave radiation is either positive or negative depending upon the value of net radiation i.e., whether it is directed towards the Earth or in the opposite direction.

Net outgoing radiation

The difference between the long wave radiation from the ground surface upwards and the downward long wave radiation from the atmosphere to the ground surface. This is also referred to as 'Nocturnal radiation'.

Net plot

The area from which yield or other characters are measured. It is also known as the net area of the plot.

Net primary productivity of an ecosystem

The total amount of chemical energy left after it has been utilized by plants for respiration.

Net pyranometer

An instrument used to measure the difference of the solar radiation falling on both sides of the horizontal surface from the areas to which the exposure is directed with this instrument.

An instrument used to measure the difference between the incoming and outgoing terrestrial radiation.

Net Pyrradiometer / Radiation balance meter
An instrument used to measure the difference of the total radiations falling on upper side and lower side of the plane surface from the solid angles.

Net radiation
To write in any form and to measure even with the best equipment, this is a complex and difficult component of solar radiation. Basically, this is the difference between the downward and upward radiation flux. When it is to say the downward flux, it is mostly the net short-wave radiation and most of the upward radiation is long wave radiation. So, to be precise, the net radiation is the net flux of total radiation passing through or available on a horizontal plane just above the ground at the moment of measurement.

Net radiometer
An instrument used to measure the net radiation.

Net solar radiation
This is the difference between the solar radiation received from the Sun and the solar radiation directed upwards. So, this is the net flux of solar radiation. This term differs from the net radiation in a way that this does not account for terrestrial component of radiation as far as expression is concerned. However, there is a lot of debate on the usage of this term as compared to net radiation, at specific instances.

Net terrestrial radiation
The difference between the downward and the upward terrestrial radiations. In other words, this is the net flux of terrestrial radiation. In some micrometeorological studies of crop plants, this term is also known as 'Terrestrial radiation balance'.

Neutral air
The condition of the air in stable equilibrium . Under this state, the air has more potential energy. If neutral air parcel is displaced, it might return to original state.

Neutral atmosphere
See 'Neutral stability'.

Neutral stability

This is the state of an atmospheric layer in which a column of unsaturated air has its environmental lapse rate of temperature equal to dry adiabatic lapse rate or a saturated air parcel has its environmental lapse rate of temperature equal to saturated adiabatic lapse rate. In neutral stability conditions, an air parcel will not experience buoyancy acceleration, if lifted vertically. So, on imposition of vertical displacement, the air parcel accepts the same without any positive and negative response.

Neutron moisture meter (depth probe)

Neutron source and detector in a cylindrical container that can be lowered down a tube in soil or other material and calibrated to indicate moisture content.

Neutron

An elementary particle with no charge in the nucleus of an atom.

Newton

The force that gives to a mass of one kilogram an acceleration of one metre per second. In M.K.S. system of units, this is the unit of force.

Newton's law of motion

The first law states that, every body continues in the state of rest or uniform motion in a straight line until an external force acts on it. The second law states that, the rate of change of momentum is proportional to the applied force and takes place in the direction in which it acts. The third law is that, for every action there is an equal and opposite reaction.

Nexrad

Acronym for Next Generation Weather Radar.

Night soil

Night soil is human excrement, solid and liquid. In India, it is directly applied to the soil to a limited extent and as town compost to a large extent. In cities which have sewage facilities, sewage water, and sludge are used directly to raise crops. On an average night-soil contains 5.5 per cent nitrogen, 4.0 per cent phosphorus (P_2O_5) and 2.0 per cent (K_2O) on oven dry basis.

Nimbo or Nimbus

A low cloud which can produce rain, snow or sleet.

Nimbo-stratus

These are lower middle level clouds and the mean height is 2 to 7 kilometres. These clouds are grey and dark in colour, thick enough to cover the Sun totally. Often gives rain, snow, sleet, etc.

Nitrification inhibitors

Are chemicals used to control nitrogen transformation in the soils, Due to their toxic effect on nitrifying bacteria, they temporarily inhibit the product of nitrate. Eg. N-serve, nitrapyrin, potassium azide etc. Application rate 0.2 to 0.6 kg/ha or active ingredient.

Nitrification

Is a two-step process in which the ammonia or ammonium is converted to nitrite and finally to nitrate and these two steps are carried out by the autotrophic micro organisms–Nitro-somonas and Nitrobacter respectively.

Nitrogen assimilation

The incorporation of nitrogen compounds into cell substances by living organisms.

Nitrogen cycle

The sequence of transformation undergone by nitrogen wherein, it is used by one organism, later liberated upon the death and decomposition of the organism, and is converted by biological means to its original state of oxidation to be reused by another organism.

Nitrogen deficiency

Deficiency during early growth stages results in yellow to yellowish green leaves, stunted and spindly growth and reduced tillering. If deficiency persists to maturity, the number of grains per head is reduced. Deficiency appears first on older leaves later entire crop eventually will appear uniformly yellow.

Nitrogen fixation

The conversion of elemental nitrogen to organic combinations, or to forms readily utilizable in biological process, by nitrogen-fixing microorganisms. When brought about by bacteria in the root nodules of leguminous plants, it is referred to as symbiotic, if by free-living microorganisms acting independently, it is referred to as nonsymbiotic or free fixation.

Nitrogen fixing plant

A plant which can assimilate and fix the free nitrogen of the atmosphere by the aid of bacteria living in the root nodules. Legumes with the associated rhizobium bacteria in the root nodules are the most important nitrogen fixing plants.

Nitrogen

Colourless, tasteless, odourless gas, which occurs in the atmosphere to the extent of 78% by volume. An essential plant nutrient, taken up by plants in the form of ammonium or nitrate ion.

Noctilucent clouds

The clouds of tiny ice particles that form approximately 75 to 90 kilometers above the earth's surface. That appear bright against a dark night sky, in blue-silver color or orange-red.

Nocturnal radiation

See 'Net outgoing radiation.

Nocturnal thunderstorms

Thunderstorms which develop after dusk.

Nodes

The points of zero displacement to which two waves arrive out of place, in a system of standing waves. This results in no vibration.

Noise pollution

Sound is a physical disturbance in a medium such as a gas, liquid or solid which can be detected by human ear. Noise is a sound of random nature and noise pollution is an unpleasant, unwanted or undesirable sound.

Noise

Random fluctuations in a parameter caused by effects other than the one being studied.

Non-selective herbicide

A material that tends to kill all plants with which it comes in contact.

Normal ploughing (plough-sole depth)

Ploughing up to a depth of 15cm.

Normal spectrum

Also known as 'Biological spectrum'. A tabulation by percentages of the plants of a crop community or region into life form classes according to Raunkiacr's classification. This classification is based on the kinds and position of the organ with respect to the soil that survives unfavourable environmental periods.

Normal

The line drawn perpendicular to a plane.

Normalize

A process in which a quantity is produced to a non-dimensional ratio. This is done by dividing a quantity by a fundamental quantity of same dimension.

Normand's theorem

The theorm expressed in the form of construction on a tephigram and also on any of the thermodynamic diagrams. The relationship is developed among temperature, thermodynamic wet bulb temperature, and dew point temperature. This relation is also helpful in certain expressions related to heat changes and the conservation of energy.

North Pacific high

A semi-permanent, subtropical area of high pressure in the North Pacific Ocean.

Norther

A storm with northerly wind which begins suddenly during the colder months in a year.

Northerly

The air in motion from northern direction.

Notillage planting

Direct planting is essentially unprepared seed beds. Weeds can be controlled by herbicides.

Nowcast

A short-term weather forecast for expected conditions with 0-2 hour validity.

Noxious weed

A weed plant especially undesirable, troublesome and difficult to control. e.g. Cynodon dactylon, Cyperus rotundus, Sorghum halepense.

Nucleation

This is a phase in which a condensed substance changes to further condensed state, i.e., into ice crystals or drops of rain.

Nuclei of condensation

The particles on which condensation of water vapour takes place and either rain or ice particles form and grow in size in the atmosphere. These particles are termed for convenience as nuclei because of their microscopic and sub-microscopic sizes. These have got strong affinity for water. In areas near seas and oceans, sodium chloride is the nuclei of condensation to a larger extent.

Nucleus seed

The original seed produced for the first time by the plant breeder. It is cent per cent pure in all genetical and physical quantities and used as a parent for the multiplication of breeder's stock seed.

Nucleus

In cloud physics and meteorology, any particle having affinity to water and on which condensation takes place. It may also be noted that in an atom the core which is positively charged is also called a nucleus. Also see 'Nuclei of condesation'.

Null hypothesis

Hypothesis of no difference; that is there is no discrepancy between observation and expectation based on some set of postulates.

Numerical forecasting

The use of numerical models, (equations of hydrodynamics) subjected to observe initial weather conditions, to forecast the weather.

Numerical model

In the models numericals are involved, i.e., the governing equations are solved by means of step by step numerical calculations. The knowledge and capability to solve equations is essential in these models.

Nurse crop

A companion crop which nourishes the main crop by way of nitrogen fixation and or adding the organic matter in to the soil, e.g. cowpea intercropped with cereals or new plantations of fruit trees.

Nutritive value index (NVI)

Intake of feed DM × digestibility of DM × efficiency of utilization of digested nutrients.

O

Oasis effect

Is the exchange of heat between a growing crop and hot air whereby air over the crop is cooled.

The vertical sensible heat energy transfer from the air to the crop canopy while flowing across the crop. The air is cooled until it reaches equilibrium with the crop canopy. In this process, a small quantity of radiant energy may also be taken by the crop from the same air under hot dry conditions.

Obscuring phenomenon

An atmospheric phenomena, which covers a part of the visible sky such that no light is seen from the point of observation. Here, the covering by clouds is excluded.

Observation

The evaluation of one or more meteorological elements, (temperature, pressure, wind, etc.,) that describe the state of the atmosphere.

Occluded front

A 'front' separating two cold air masses. This process should be as a result of occlusion and also when these two cold air masses are brought into contact with each other. Also see 'Occlusion'.

Occlusion

A 'front' forming when a cold front overtakes a warm front. This happens when a depression is developing. When occlusion takes place, the warm sector is potentially lifted off the ground.

Ocean current

The enormous movement of ocean water from an area of warm temperature to another area of relatively colder temperaure or vice versa. Basically, like air current, these ocean currents also move from one place to another in order to equalize the temperature.

Oceanography

The science dealing with all aspects of the ocean which includes the physical, chemical, animal, plant life, etc. As far as meteorology is concerned, ony physical conditions are studied in relation to ocean.

Official seed sample

The samples which are drawn by a seed control or seed law enforcement officer. These are submitted to the laboratory to determine if a seed lot being offered for sale, meets the requirements of the seed act.

Offtype

Plants or seeds deviating significantly from the characteristics of a variety as described by the breeder in any observable respect.

Ohm

A unit of electrical resistance.

Ohm's law

The ratio of the potential difference between the ends of a conductor and the current flowing in the conductor is constant. In other words, the current in an electrical circuit in amperes is equal to the ratio of electromotive force and resistance, for which volts and ohms are the respective units. An ohm is the derived S.I. unit of resistance. This law is derived by George Ohm.

Oil (plant)

A slippery viscous liquid or fat like substances produced and stored in the seeds of plant species such as sunflower, groundnut, soyabean etc.

Okta

An okta is a unit for estimating the cloud coverage of the sky and is equal to one eighteenth of the total exposed hemisphere of the sky.

Olericulture

A branch of horticulture which deals with cultivation of vegetables.

Oligotrpoic

The term is used to describe a body of water (a lake, tank, or a pond) with a poor supply of nutrients and a low rate of formation of organic matter by photosynthesis.

Omega block

A warm high aloft which has become displaced and is on the polar ward side of the jet stream.

On-farm research and development

Agronomic and socio-economic studies conducted on the farms with farmer's active participation. The goals are to develop improved cropping system technologies and to devise ways to combine these technologies with farmer's knowledge and skill to efficiently utilize the available farm resources.

Opaque bodies

Bodies that do not permit the passage of light.

Open pan evaporimeter

An instrument used to measure the rate of evaporation from free water surface. (This term is the short form of the United States Weather Bureau Class. An open evaporimeter).

Operational research project (ORP)

Is a transfer of new technology. It is an integrated approach launched by ICAR in cooperation with local agencies, voluntary organizations, agricultural universities, state agricultural departments to test, adopt and modify, if necessary, (i) research findings and made them suitable for large-scale adoption by farmers, (ii) to find out the constraints that impede acceptance and release of such findings and (iii) to find out the profitability of such findings.

Ophelion

The Earth is little farther from the Sun on about July 4th each year, making the distance in between as 152 million kilometers. This position is called ophelion. Also see 'Perihelion'.

Opportunity cropping

The practice of placing primary emphasis on the use of stored soil moisture while determining whether or not to establish a crop.

Optical air mass

A measure of length of the distance travelled by light rays from an extraterrestrial celestial body or a heavenly body through the atmosphere upto the mean sea level.

Optical density

The disagreement of a transparent material to the passage of light. So, the index of refraction of the material is the measure of optical density of the same material.

Optical depth

A means of the approximate distance in an absorbing medium for a monochromatic beam. This is the logarithmic ratio of the radiance at one point to the radiance after transmission.

Optical pyrometer

An instrument used to measure very high temperatures. The intensity of radiation from a black body is compared with the emission of the same wavelength from a standard map.

Optical thickness

A measure of the attenuation of solar radiation by the atmosphere due to different processes like scattering, absorption, etc. The optical thickness and attenuation are directly proportional to each other.

Optimum climatic conditions

The range of weather and climatic conditions which are most favourable for a crop to complete its life cycle.

Optimum tillage

A system of tillage contributing to maximum net return for a crop under given field conditions

Orbit

The elliptical path of revolution of one heavenly body around the Sun or another body. This exists as a result of mutual gravitational attraction.

Order of magnitude

The degree of the highest derivative in a differential equation is 'order' and the 'order of magnitude' is a typical magnitude of a quantity to the nearest integral power of ten.

Ordinary agricultural meteorological station

A meteorological station in which data are collected not only on weather elements but also on the condition of crops including other biological information. The data thus collected are used in conducting research on specific agrometeorological problems of the same area.

Organic farming (biological husbandary)

It is an agricultural production system which avoids or largely excludes the use of synthetically compounded fertilizers, pesticides, growth regulators and livestock feed additives. To the maximum extent feasible organic farming systems rely upon crop rotations, crop residues, animal manures, legumes, green manures, mineral-bearing rocks and aspects of biological pest control to maintain soil productivity and tilth to supply plant nutrients and to control insects weeds and other pests.

Organic manures

Carbonaceous materials mainly of vegetables or animal origin added to the soil specifically for the nutrition of plants.

Organic matter

It includes paint and animal resides of various stages of decomposition in soil, cells and tissues of soil organisms, substances synthesized by soil microbial activity. Plant leaves, stem, straw, husk, roots, stubbles or any ramnents of crops of plant species and organic manures including FYM are the cheap source of organic matter.

Organic nitrogen (soil)

Nitrogen in the form of organic compounds e.g. amino acids, amino sugar, purine and pyrimidine bases.

Organic soil

Soils containing organic matter in sufficient quantities to dominate the soil characteristics. Frequently all soils containing 20 per cent or more organic matter by weight are arbitrarily designated as organic soils.

Organism

A self regulating and self perpetuating physico-chemical entity which is in perfect balance with the environment.

Oriented tillage

Tillage operations which are oriented in specific paths or directions with respect to the sun, prevailing winds, previous tillage actions or field base lines.

Orogenic

See 'Orographic precipitation' and 'Orographic lift'.

Orographic lift

The lift caused to an air mass when it is forced to ascend over the mountain range.

Orographic precipitation

Of the different types of precipitations, the orographic precipitation occur due to the presence of mountain barriers and the assent of moist air over it. When moisture laden current of air strikes a mountain range, it is forced upwards causing condensation and precipitation , especially on the windward side of the mountain barrier.

Orographic

Any phenomenon pertaining to the hills and mountains and same as 'Orogenic'.

Osmosis

The diffusion of water or any other solvent across a semipermeable membrane from a region of its higher concentration to a region of lower concentration.

The diffusion of a solvent (especially water) across a differentially permeable membrane separating two solutions or separating a solution and a solvent. The concentration of salts in a solution is called osmotic concentration.

Osmotic pressure
An index of potential maximum pressure which can be developed in a solution as a result of osmosis.

Out crop
The actual edge of the inclined stratum at the surface ground.

Outer crop/guard crop/border crop
The crops which are grown around the field boundaries in narrow strips with twin objectives of protecting the main crop from stray cattle and producing livestock feed and/or seed are called out or guard crops e.g. Sesbania or Leucaena on boundaries of field/plantation crops and castor around spring planted sugarcane.

Outflow
The outward flow of air from a system.

Outlet channel
A channel which carries water out or away from a reservoir, lake or other water body.

Oven dry soil
Soil is considered to be oven dry when it has reached equilibrium with the vapour pressure from an oven at 105 ^{o}C. The oven dry condition corresponds to a relative humidity of approximately 0% of pH near 7.

Over irrigation
The application of more water than is necessary for the needs of vegetation, resulting in loss of water through seepage and leaching.

Over stocking (animal)
Stocking animals beyond the safe grazing capacity of a pasture.

Overcast sky

The sky is used to be overcast when the existing cover of the clouds is equal to 3, 4 or 5 oktas.

Over running

A situation in which a relatively warm air mass is forced above a cooler air mass of greater density. Resulting cloudiness, cool temperatures, and steady precipitation.

Over seeding

It refer to seeding of a legume/grass into an existing pasture in order to improve the productivity and quality of herbage.

Ozone

An allotropic form of oxygen, containing three atoms in the molecule and is formed in the atmosphere by the action of solar radiation on oxygen gas.

Ozonosphere

From the surface of the Earth in between 30 and 60 kilometers a large quantity of ozone is present in a layer in the mesosphare. This ozone layer is called the ozonosphere which is warm and absorbs the ultra-violet, cosmic rays, etc., of the solar radiation and enables plant, animal and human life to exist on the Earth.

P

p.t.h

This is an abbreviation used in radiosonde observations to indicate the pressure, temperature and humidity.

Paddy weeder

An equipment for interculture used in paddy cultivation. It is used for uprooting weeds and burying them in puddle soil between rows of standing crop. It improves the aeration of the soil.

Pan coefficient (k)

Ratio between reference evapo-transpiration (ET^O) and water loss by evaporation from an open water surface of a pan (E).

Panicle initiation stage

Begins when the primordial of panicle has differentiated and becomes visible.

Parabola

A geometric curve traced out by a point which moves, so that its distance from a fixed point, the focus is equidistant from a fixed straight line.

Parallax

The difference in direction or the apparent change of relative position of two objects due to a change in position of the observer.

Parallel circuit

In electricity, the complete path traversed by an electric current is called electrical circuit. So, in parallel circuit the electric current is divided between two or among more conductors.

Parboiling (rice)

Quite often the paddy is subjected to parboiling treatment before milling and it involves soaking the paddy in water, steaming it in hot water and drying it in the sun. Parboiling makes husk-removal easier, toughens the grain which results in reduced breakage during milling and polishing and improves the keeping quality of rice. Also parboiled rice retains more nutrients while processing and cooking.

Parcel

The volume of air small enough to contain uniform distribution of its meteorological properties and large enough to remain relatively self-contained and respond to all meteorological processes.

Parshal flume

A calibrated device for measuring the flow of water in open conditions. It consists of a contracting length, a throat and an expanding length.

Partial correlation

Partial correlation coefficients measure that strength of the linear relationship between one dependent variable and one of several independent variables after controlling the effects of the other independent variables by holding them statistically constant.

Partial obscuration

If $1/8^{th}$ or more of the sky (not full sky) is hidden by any surface-based phenomena in the atmosphere, excluding precipitation.

Partial pressure

The pressure exerted by a particular component of a gaseous mixture if other components of the same mixture are removed without change in the volume.

Particle density (soil)

Particle density is defined as the ratio of mass of solids (oven dried soil) to the volume of the solids alone and is expressed as g\cm^3.

Particle model

Different models are being developed for various phenomena over a period of time, as science is advancing and the behaviour of light is also described in several ways. Quantum theory is the one which believes that, the light energy exists in discrete units called quanta. The wave theory believes that, the energy exists in particles with elastic displacement and exhibiting periodic disturbances. So, when the behaviour of light is expressed in particle properties and a model thus arrived at described as particle model.

Partly cloudy

The state of the weather when the clouds are conspicuously present.

Pascal

The S.I. unit of pressure equivalent to one Newton per metre per metre.

Pascal's law

The law which states that, the pressure applied anywhere to an enclosed body of fluid is transmitted equally in all directions and acts with equal force on equal areas. The force per unit area being uniform throughout, this pressure acts at right angles to every portion of the surface of the container.

Past weather

The main characteristics of the weather that existed at an observatory, station, or location, in any required period of time that elapsed. This information is represented in a coded form as per the specifications of the IMD in India.

Pasture land

An area of land covered with grass or other herbaceous forage plants used for grazing animals.

Pasture renovation

Improvement of a pasture by tillage, seeding, fertilization and some times by liming also.

Pasture

The term pasture is applied to a grazed plant community usually composed of several species often of diverse botanical type.

Peak gust

The highest instantaneous wind speed observed or recorded.

Peat (soil)

Unconsolidated soil material consisting largely of undecomposed, or only slightly decomposed organic matter accumulated under conditions of excessive moisture.

Pedal equations

The relation between the distance of any point on a curve from the origin (or pole), and the length of the perpendicular from the origin (or pole) to the tangent at the point.

Pedology

The science dealing with the laws of origin, formation and geographic distribution of the soil as a body in the nature.

Pelleted seed

Layer of particular matter around the seed, pelleting is usually made from finely pulverized limestone or mineral phosphate.

Penumbra

Half or partial shadow of an object. This is formed when an object in the path of rays from a large source of light obstructs and reduces a part of the light.

Percent germination

The proportion of the number of seeds which have produced seedlings classified as normal under the conditions and period specified.

Percent possible sunshine

The ratio of time of actual bright (unimpended) sunshine to the time during which sunshine is possible.

Perched water

The body of free ground water in a zone of saturation, separated from an underlying body of ground water by an unsaturated layer of material.

Perched watertable

The upper limit or surface of a small body of water above the main water-table. The water is retained in its elevated position by an impervious stratum and may form a limited source of water supply.

Perennial

Continuing or lasting for more than two years. A plant that dies back seasonally and produces new growth again form a perennial part.

Perfect gas equation

The product of pressure and volume is equal to the product of universal gas constant and temperature. A perfect gas is an ideal gas which would obey the gas laws perfectly.

Perihelion

The point of the earth's orbit that is nearest to the sun.

This is a situation in which the Earth is a little closer to the Sun on about January 3rd each year making the distance in between as 147 million kilometers. Also see 'Ophelion'.

Period

In a battery of continuous events, the interval between two consecutive stages and usually this quantity if expressed as a function of time.

Periodicity

The repeated occurrence of events at fairly frequent and regular intervals.

Permanent cropping system

An agricultural system in which crop plants are grown continually on the same plots of land, either several crops in rotation or a single crop for several years; disturbance of microclimate and of water relations is permanent; nutrients and organic matter are replaced by inputs from out-side.

Permanent pasture

Pasture of perennial or self seeding annual plants maintained for several years for grazing.

Permanent wilting point

The moisture content in percentage of a soil at which nearly all plants wilt and do not recover in a humid dark chamber unless water is added from an outside source. This is the lower limit of available moisture range for plant growth. Below the wilting point extraction of moisture continues for some time but growth ceases completely. The force with which moisture is held by the soil at this point corresponds to 15 atmospheres.

Permeability rate

The rate at which water moves through a soil under a standard pressure gradient, usually defined as amount of water traversing per sq.cm of soil surface under a pressure gradient of one dyne per cm.

Permeability

The property of a porous medium to transmit fluids. It is a broad term and can be further specified as hydraulic conductivity and intrinsic permeability. 'K' in Darcy's law and represents the rate of flow of water in a porous medium under a unit hydraulic gradient. It changes with the quality and quantity of water but the value of intrinsic permeability, the property of porous medium remains constant for any fluid.

Persistent herbicide

A herbicide which, when applied at the recommended rate will harm susceptible crop (s) planted in normal rotation after harvesting the treated crop or which interferes with regrowth of native vegetation in non-crop sites.

pF

Is the logarithm of height in cm of a water column that represents total stress with which water is held by a soil.

258

pH of soil

The negative logarithm of the hydrogen-ion activity of a soil. The degree of acidity (or alkalinity) of a soil as determined by means of a glass quinhydrone or other suitable electrode or indicator at a specified moisture content or soil to water ratio and expressed in terms of the pH scale, from 0 to 14 invented by Sorenson.

Phased planting/(staggered planting)

Is planting/sowing of a crop spread over an optimum period of planting either to minimize risks or to use labour/machinery more effectively or to minimize competition (in intercropping) or to prolong the period of supply to the market or the factories.

Phenology

The study of the timing of recurring biological events, the causes of their timing with regard to biotic and abiotic factors and the inter-relation among the phases of the same or different species.

A study of the developmental timing of organisms in relation to their environment.

Phenophases (growth stages)

- A period in the developmental timing of a crop plant in relation to its environment.
- The stages of development in the life cycle of plant in relation to the climatic conditions. The important phenophases in cereal crop plants are emergence, tillering/branching, heading, flowering, anthesis etc.

Phenotype

Physical expression or outward appearance of a genotype which is subject to change with the environment.

The expression of the characteristics of a crop plant as determined by the interaction of both genetic constitution and environment.

Phoresis

The passage of particles under the influence of a thermal field containing energy.

Phosphorescent bodies

The bodies for which luminescence persists after removal of the existing source. These are also called as 'after glow bodies'.

Photochemical

The chemical reactions which take place because of the absorption or emission of solar radiation. The photosynthetic process is one of the classical examples for the photochemical reactions, in crop plants.

Photochemistry

The study of chemical reactions of the substances which are initiated, assisted or accelerated by exposure to photons of electromagnetic radiation.

Photodissociation

In meteorology, the molecular dissociation caused by the incidence and absorption of the photons of solar radiation.

Photoelectric cell

The process by which the electromagnetic radiation incident on a solid, liquid or gas releases electrical charge which is detectable in electric field is called as photo-electricity. This does not occur by wave nature but by quantum process only. So, photoelectric cell is a device for causing the release of electrons, in photoelectricity.

Photoelectric effect

The effect caused by the emission of electrons from a substance on irradiation with light.

Photoelectron

The electron released from a substance after absorbing the energy from light.

Photoionisation

Ionisation caused by the absorption of photons of electromagnetic radiation.

Photokinesis

The undirected motion of many lower plants in response of light.

Photometer

An instrument used to measure the intensity of light source, light distribution, luminous flux, colour, etc.

Photometry

The measurement of properties of visible radiation. In the measurements the eye of the human being is considered as primary sensor, which covers 400 to 700 microns in the solar band.

Photomorphogenesis

The responses of plant to light in which a distinct and complex change occurs in the pattern of differentiation.

Photon

A quantum associated with electromagnetic energy and the fastest moving particle in the universe known, as on date.

Photonasty

The response of an organ or whole of a plant to the incident diffused light.

Photoperiod

The duration in which the plants are exposed to light during a day. This is expressed in hours in a day or cumulative hours.

Photoperiodism

The response of crop plants to the relative duration of sunlight and darkness. Also see 'Short day plants and long day plants'.

Photorespiration

The process of respiration that occurs in plants in the presence of light.

Photosphere

The visible and intensely luminous upper layer of the convection zone of the Sun which has sharply defined edges.

Photosynthesis

The assimilation of water and carbon dioxide in the chlorophyll containing tissues of plants, when exposed to light. In this process, the carbohydrates are formed leaving the oxygen as a byproduct. From the view point of an agrometeorologist, this synthesis comprises (a) the diffusion process; (b) the photochemical process; and (c) the biochemical process.

Photosynthetic efficiency

The percentage of total solar energy assimilated by plants. The energy fixed by plants expressed as percentage of the incoming solar energy.

Photosynthetic photon flux density

The number of photons in the visible band of the solar spectrum incident per unit time on a unit surface.

Photosynthetic pigments

The plant pigments responsible for the capture of light energy during light reactions of photosynthesis. The green pigment chlorophyll is the principal light receptor, absorbing blue and red light. However, the carotenoids and various other pigments also absorb light energy and pass this on to the chlorophyll molecules.

Photosynthetic quotient (PQ)

The ratio of volume of O_2 evolved to the volume of CO_2 absorbed in photosynthesis,   or $PQ = O_2/CO_2$.

Photosynthetically active radiation

The radiation that exists in the wave band between 0.4 to 0.7 microns. Because of the importance of this particular band with regard to photo-synthetic process in green crop canopies and to distinguish it from other brands, this term is commonly being used by micrometeorologists in general and all agrometeorologists in particular.

Phototropic plants

The plants which can obtain energy from the sunlight.

Phototropic

The curvature turning of an organ of a crop plant towards the light of greater intensity when unequally illuminated.

Phototropism

A phenomenon, in which the visible part of the electromagnetic radiation influence the orientation of the shoots of a crop. The phototropism is of two types viz., the positive and negative. When the shoots turn towards the light on incidence it is positive phototropism, and the opposite is the negative phototropism.

Phreatic rice lands (upland/low land rainfed)

Whether they are naturally sloping or flat, may or may not be bunded and free ground water is present within the root zone of rice plant during the growing season (It defers from pluvial with regard to water table depths.)

Physical climatology

A branch of climatology which studies the temporal and spatial variations in heat and moisture exchanges, and also the movement of air in the atmosphere.

Physical colour

A colour of a substance as seen by an eye, because of the emission of wavelengths from waveband of electromagnetic spectrum. This term is restricted to a class of differences perceived at equal luminance.

Physiological drought

- Non availability of water to the plant due to unfavourable physiological conditions, such as water logging, soil salinity, low temperature etc.
- The drought experienced by a plant, in which it wilts or suffers from insufficient water although its surroundings contain sufficient water.

Phytoclimatology

Study of microclimate in relation to plants.

Piche evaporimeter

A filter porous paper atmometer used to measure the amount of water evaporated.

Piezo electric effect

The electricity or electrical impulses resulting from the application of mechanical pressure on a dielectric crystal. For converting a mechanical strain into an electrical signal piezo electrical materials are used in transducers. The essential situation for this effect to happen is, the absence of symmetry in the crystal structure.

Pigment

Any material which imparts a physical colour to a substance under vision.

Pilot balloon

A balloon of relatively small size used to obtain the data on winds.

Pilot Report

A report of in-flight weather by an aircraft pilot.

Placements (fertilizer)

Inserting, drilling or placing the fertilizer below the soil surface by means of any tool, implement or equipment at desired depth to supply plant nutrients to crop before sowing or in the standing crop.

Plan position indicator

A radar indicator scope displaying range and azimuth of targets in polar coordinates.

Planck's law of radiation

The energy of electromagnetic radiation is composed of discrete particles called quanta and the magnitude of which is given by multiplying the Planck's constant with the frequency of the radiation. The Planck's curve describes the amount of energy radiated as a function of the wavelength of a black body.

Plane parallel atmosphere

This is mainly a hypothetical phenomenon, in which it is assumed that in the atmosphere, the radiation properties are constant in parallel lines.

Planetary boundary layer

The turbulent layer of the atmosphere next to the surface of the Earth, which is several kilometers in depth.

The turbulent layer of the atmosphere next to the Earth's surface in which skin friction effects are predominant and significant. This is several hundreds of metres in depth.

Plant

A living member of any vegetables or cultivated species. A plant may perform fully or partially all the physiological processes.

Plant geometry

The pattern of distribution of plants over the ground or the shape of the area available to the individual plant.

Plant physiology

The study of mechanisms or processes in plants and interpretation of plant behaviour in relation to physical and chemical laws.

A branch of agricultural sciences, that deals with the functions and processes carried on by crop plants.

Plasma

This term has been used with different meanings in different branches of science. In agrometeorology particularly in micrometeorological studies, this is used to indicate a gas with ionized particles in sufficient density and which is a good electrical conductor.

Plastic mulch

Thin polyethylene film which may be transparent or black that is used as a mulch for field crops, benefits by moisture retention, increased soil temperature and weed control.

Plot (Experimental)

An experimental unit to which a treatment is applied (It may be single plant, small area of soil containing several plants, a strip through a field or a whole field).

Plough sole placement

A method of fertilizer application in which fertilizer is placed in a continuous band in the bottom of the furrow, in the process of ploughing each band is covered as the next furrow is turned.

Plough sole

A compacted layer at the bottom of the zone of ploughing. Plough, before the crops are sown.

Ploughing

Field operations carried out with the help of tractor drawn or bullock drawn implements.

Plow/Plough Wind

The spreading downdraft and strong straight-line winds preceding a thunderstorm.

Plumule

The major young bud of the embryo within a seed from which aerial portions of the plant develop. It usually occurs at the tip of the epicotyl, the part of the embryonic plant axis above the cotyledons.

Pluvial rice lands (rainfed uplands)

Lands which are well drained without free ground water within the rooting zone of the rice plant and are generally unbunded.

Pluviograph

A rain gauge which records the precipitation on a calibrated chart of the rotating drum, fixed in a chamber. Both the intensity and duration of rainfall can be obtained with the help of this instrument.

Pluvioscope

An instrument from which the form and duration of precipitation can be observed.

Poisson distribution

Discrete probability mass function involving events which take place rarely or relatively infrequently when only a small sub-interval of time is being considered.

Poisson's equation

Equation relating initial and final absolute temperatures and pressures of an ideal gas undergoing dry adiabatic compression or expansion.

Polar air mass

A high latitude air mass. These are three different polar air masses viz., winter time continental, summer time continental and winter time maritime polar.

Polar climates

The summerless cold climates founding polar regions and snow is the only form of precipitation. The mean warmest monthly temperature is below 10 degrees centigrade. The natural vegetation includes lichens, masses, etc.

Polar front

Any 'front' that develops in areas of the world where air masses have higher temperature differences. In general, the middle latitude fronted zone separating air flowing from tropical and polar source regions is a 'polar front'. The most important frontal zone is the Arctic polar front which has its full development during the winter. Another version for the genesis of polar fronts is, when the polar continental and the tropical maritime air masses coming in close proximity. Likewise, there are other reasons also. However all these are favoured sites of cyclogenesis.

Polar Jet

The jet which is the boundary between the polar air and the subtropical air.

Polarised light

Light waves vibrating in a single plane rather than in all directions.
    Light whose vibrations are in one plane only.

Poles/Polar

The poles are the geographic point at 90 degrees latitude north and south on the earth's surface. The polar region is that area between 60° and 90° latitude, both north and south of the equator.

Pollutant

Particles, gases, liquid aerosols etc., in the atmosphere which have an undesirable effect on humans surroundings.

Pollution

An undesirable change in the physical, chemical or biological characteristics of the natural environment brought about by the activities of man. The effects may be on soil, water, environment or on human beings. The pollutants are classified into two. They are biodegradable (eg. Sewage) and non-biodegradable (e.g. pesticides).

Polynutrient fertilizer

A fertilizer containing more than one major plant nutrient. It is synonymous for multiple nutrient material and complex fertilizers.

Population density

In micrometeorological field studies, the number of individual plants that are existing in a unit area usually in an area of a square metre.

Pores (macro)

When diameter is 100 microns or more, main functions are to maintain air circulation and drainage of soil.

Pores (meso)

Pore size diameter ranges between 30 to 100 microns and helps in capillary water conduction or movement.

Pores (micro)

Diameter of pores ranges from 3 to 30 microns; these play important role in water retention (slow capillary flow).

Porosity (pore-space)

Is the percentage of soil volume not occupied by the soil particles.

Calculated by the formula $(N) = \dfrac{P.d - b.d}{p.d} \times 100$

where

N   = Porosity
P.d = Particle density
b.d = Bulk density

Positive vorticity advection

The advection of higher values of vorticity into an area.

Positively charged body

This is the property of a body that has lost the elementary particles called electrons which are said to be negatively charged with electricity.

Post harvest tillage

The cultivation of the land immediately after harvesting at physiological maturity to aid in the preparation of the land for the next crop.

Post emergence treatment

Treatment made after emergence of specified weed or crop.

Potassium absorption ratio (PAR)

A ratio for soil extracts and irrigation waters used to express the relative activity of potassium ions in exchange reactions with soil.

Potassium deficiency

Symptoms of mild deficiency are dark green leaves, low tillering and stunting. Severe deficiency results in yellowish-orange to yellowish-brown discoloration starting at the tip of older leaf blades and gradually moving towards the base. Necrotic spots may appear on the leaf blade. Grain size and weight may be reduced. Potassium deficiency usually occurs in coarse, sandy, or peat soils and in soils rich in potassium fixing clays.

Potassium nitrate (KNO$_3$)

A nitrogen-potassium fertilizer containing 13 percent N and 44 per cent K$_2$O.

Potential crop production

The total dry matter production of a crop surface during its entire growth period when it is optimally supplied with water and all essential nutrients, and grows without interference from weeds, pests and diseases.

Potential crop yield

Yield of a crop obtained at research stations under most ideal environment.

Potential difference

This is expressed, basically, in terms of energy. The work performed to move a unit charge from one point to another is the potential difference between these two points. In other words, if two points have different electrical potentials, it is to be stated that a potential difference exists between them. An instrument used to measure this potential difference is potentiometer.

Potential energy

The energy of a body or a system of bodies due to its position in the field of force such as gravitational, electric, spring, etc. This is the product of the mass of the body, its height above the ground level, and acceleration due to gravity.

Potential evaporation

Similar to 'Evaporation'.

Potential evaportranspiration

The water transpired from a uniform, short, green, actively growing vegetation when water supply is unlimited. This is expressed as the rate of latent heat transfer per unit area or an equivalent depth of water. The units are millimeters.

Potential soil water

Is the amount of work that must be done per unit quantity of pure water in order to transport reversibly and isothermally an infinitesimal quantity of water from a pool of pure water at a specified elevation and atmospheric pressure to the soil water at the point under consideration.

Potential temperature

The temperature of an air parcel would have been, if brought adiabatically to a standard pressure of one thousand millibars. So, this is the temperature of an air parcel after dry adiabatic compression.

Potential total

Is the sum of all potentials acting on water in an equilibrium system. It is the amount of work that unit quantity of water in an equilibrium soil-water (plant-water) system is capable of doing when it moves to a pool of pure free water at the same temperature located at a reference level and subject to atmospheric pressure.

Potential water

Is the difference between chemical potential of water in an equilibrium system and the chemical potential of water at the same temperature in the reference state. Thus it is the sum of all components of chemical potentials for isothermal conditions i.e. it is the sum of the pressure potential, the matric potential and solute potential. The water potential is the amount of work that a unit quantity of water in an equilibrium soil-water (or plant-water) system is capable of doing when it moves to a pool of pure water in the reference state at the same temperature.

Potometer

An instrument used to measure the transpiration as well as absorption of water by plants.

Pound

The unit of weight in British system, equal to 0.453 kilograms. This is used as a unit of force and is the force of attraction of the Earth upon a mass of one pound.

Poundal
Like pound this is also a unit of force, equal to one pound divided by thirty two. Also see 'Pound'.

Pounds per square inch (PSI)
A unit for measuring pressure, in which one PSI equals the pressure resulting from a force of one pound acting over an area of one square inch.

Power law profile
In crop canopies, the aerodynamics at micro level are very subtle to record and the logarithmic wind velocity profiles are commonly used to find out the relative gradients of wind and its transfer processes. As an alternative to this logarithmic velocity profile, the power law profile can also be used for equally accurate readings and observations.

Power tiller
A walking type of tractor. This tractor is usually fitted with two wheels only. The direction of travel and its control for field operations is performed by the operator walking behind the tractor.

Power
Time rate of doing work. The derived S.I. unit is watt.

Precipitable water
The quantity of water which must be obtained when the water vapour in a given volume of atmosphere is condensed and precipitated.

Precipitation effectiveness ratio
The total amount of precipitation for a certain period of time divided by the total amount of evaporation, both in liquid form.

Precipitation rose
A radial diagram expressing the amounts of precipitation by months or other intervals of time.

Precipitation
The falling of any condensed moisture from the cloud to the ground, depending upon temperature of the atmosphere. It may occur as rain, snow, hail, sleet, etc.

Precision

It denotes relative or apparent nearness to the truth.

It is the ability of the instrument to reproduce a set of readings within a given accuracy.

Pre-emergence treatment (herbicides)

Herbicides applied after seeding but prior to crop or weed emergence.

Pre-frontal squall Line

A sequence of thunderstorms that precedes an advancing cold front.

Pre-frontal trough

The enlarged area of relatively low pressure preceding a cold front.

Pregerminated seed

The procedure of germinating seed before sowing, usually until the radicle just emerges, as practiced in direct seeded rice.

Preparatory tillage

Various tillage operations carried out on the farm before the crops are sown, planted or transplanted to make a fine seedbed.

Preplanting treatment (herbicides)

Treatment applied after the soil has been prepared but before seeding/planting.

Present weather / Current weather

The prevailing status of different weather elements which ultimately gives the condition of the weather at any observatory, location, place, etc. This information should be represented as per the IMD guidelines in India.

Pressure altimeter

An aneroid barometer used to measure the pressure, with a scale graduated in altitude. In ordinary barometers, common pressure units are indicated, which is not the case in this barometer.

Pressure altitude

The altitude in standard atmosphere at which a given pressure is observed.

Pressure change

The difference between the values of barometric pressure at the beginning and ending of a specified interval of time.

Pressure characteristic

The pattern of the pressure change (fall, rise, steady etc.,) during the specified period of time,

Pressure gradient force

The net force on an air parcel in the atmosphere, arising from its location in a situation of pressure gradient.

Pressure gradient

The rate of increase or decrease of pressure in unit distance at a particular time.

Pressure jump

An instantaneous and sudden increase in station pressure.

Pressure tendency (barometric tendency)

The rate of change of pressure at the surface of the Earth.

Pressure tube anemometer

An instrument used to measure the wind speed by employing a pilot tube.

Pressure

The force exerted per unit area of a surface determined by the bombardment of air molecules on the same surface. Expressed by one Newton per metre per metre or one pound per square inch or one atmosphere. The S.I. unit of pressure is Newton per metre per metre which is also called as pascal.

Prevailing visibility

The visibility conditions at the observation station.

Prevailing wind

At any place, a wind is said to be a prevailing wind, when it has a clearly higher frequency in one direction than the winds of other direction.

Primary colours

Red , green and bluish-violet, the combination of which form white light. In some cases, the pigment colours red, yellow and blue which cannot be imitated by mixing any other pigment colours are also called by this term.

Prey crop (smother crop)

The crop which is grown with the purpose of eliminating any undesirable plant through physical or its allelopathic effect. eg:- Cowpea

Primary radiation standards

Very high precision and reliable radiation instruments which are maintained as reference standards at the world radiation center, the headquarters of which is Davos, Switzerland.

Primary tillage

Tillage operations which constitute the initial major soil working operation. It is normally designed to reduce soil strength, cover plant materials and rearrange aggregates.

Principal focus

The point at which converging rays of light meet is called real focus. The point to which parallel rays of light are brought by a mirror, lens, etc., is principal focus.

Probability

The proportion of times in which an event occurs in an infinitely large and hypothetical series cases, each capable of producing the event.

Productivity of an ecosystem

This is the rate at which the producers convert Sun's radiant energy into chemical energy for storage in an area under study.

Profile

A graphic representation of the magnitude of a quantity of a weather element as a function of height, in agrometeorological measurements. For example, a wind profile over a wheat crop or a temperature profile of soil under different depths, in a crop canopy, etc. Also see 'Temperature profile'.

Profiler

A Doppler radar which typically measures both wind speed and direction

Prognostic chart

A chart of forecast predictions of weather elements.

Prominencies

These are some of the various indices of solar activity, which are tenuous filaments ejected upwards and seen from solar limb of the solar atmosphere.

Proton

A positively charged particle in the nucleus of an atom with mass 1836.12 times greater than that of an electron. A proton differs a lot from a photon. The smallest indivisible quantum of electromagnetic radiation which has zero rest mass is a photon. Photons are generated, when a particle possessing an electric charge changes its momentum in collisions between nuclei and electrons.

Psychrometer

An instrument having the combination of wet bulb and dry bulb thermometers fixed on a wooden piece. This is used to measure the water vapour content of the atmosphere by comparative readings of these two thermometers.

Psychrometric chart

A monograph from which the relative humidity and dew point can be obtained graphically from wet and dry bulb thermometer readings. This is commonly used in most of the agrometeorological observatories and a semi-skilled observer is trained to record with full accuracy.

Psychrometric constant

It is the ratio of specific heat of air to the latent heat of evaporation of water, The measurement of humidity of the atmosphere is psychrometry. A psychrometric constant is the ratio of the specific heat of air to the latent heat of evaporation of water. This is used in Bowen's ratio concept of radiation energy balance studies and other evapotranspiration processes, the value of which varies from place to place.

Puddle

A compact mass of earth, soil, clay or mixture of two or more of them which has been compacted through the addition of water and rolling and trampling and made more or less impervious.

Puddled soil

Dense massive soil artificially compacted when wet and having no regular structure. The condition commonly results from the tillage of a clayey soil when it is wet.

Puddler

- An impliment used for preparation of paddy fields with standing water after initial ploughing. It breaks up clods and churns the soil. The main purpose of puddling is to reduce leaching of water.
- Making a soil impermeable by manipulating and compacting it in standing water, which reduces its apparent specific volume thus facilitates transplanting.

Pulse

A brief burst of an electromagnetic radiation emitted by the radar.

Punasa

The name given to crop season in Andhra Pradesh which extends from April – May to August – September.

Pure seed

The seed true to its kind or variety.

Pyranometer

An instrument used to measure the global solar radiation.

Pyrgeometer

An instrument used to measure the intensity of net long wave radiation (effective terrestrial radiation) at the ambient air temperature.

Pyrheliometer

An instrument commonly seen in most of the meteorological observatories, which is used to measure the intensity of direct solar radiation at normal incidence. This is considered as a passive receiver of electromagnetic energy. The common types are compensation pyrheliometer, and thermoelectric pyrheliometer.

Pyrradiometer

An instrument used to measure both incoming solar radiation and outgoing long wave terrestrial radiation.

Pyru

The crop season in Andhra Pradesh from November to March.

Q

Q$_{10}$ rule

The rate of response of a process in crop plants is often doubled or more for each increase of 10 °C of temperature within certain limits. This is also known as Vont Hoff's rule. This is only a general rule which may hold good only under certain environmental conditions, and to a few crop varieties.

Quadratic response

The output goes on increasing upto a certain level of input and then starts decreasing when input is increased further. In such cases the response is said to be quadratic and is given by, $y = a + bx - cx^2$

Quality of light

The wavelength composition of light.

Quality seed

Good quality seeds, generally true to species or variety, have capacity for high germination, are free from diseases, insects, weed seeds, inert matter, other crop seeds and extraneous material.

Quadruple cropping

Growing four crops a year in sequence.

Quantitative precipitation forecast
A forecast of rainfall, snowfall or liquid equivalent of snowfall.

Quantized
The light energy exists only in integral multiples of a basic unit like electric charge, etc. A quantity like this is said to be quantized, if in accordance with quantum mechanics, it can only have certain discrete values, each of which is called a quantum.

Quantum flux density
The number of quanta from a source, incident on a unit area of a small surface situated in a perpendicular direction.

Quantum irradiance
The number of quanta from a source incident on a unit area of a surface. This differs from the 'quantum flux density' as far as direction is concerned.

Quantum mechanics
The dynamics of atomized objects are described in different methods. In quantum mechanics, such dynamics are expressed or described in mathematical methods. The version of quantum mechanics is that the heat radiation from a black body is emitted in discrete quanta of energy, the magnitude of which are given by the product of the frequency of the radiation and an universal constant known as Planck's constant. Later on, this has been developed into a specialized form known as 'wave mechanics'.

Quantum theory
There are two theories in practice with reference to electromagnetic radiation. 1. Wave theory; and 2. Quantum theory. As per the quantum theory the radiant energy of a light beam is composed of quanta and the energy is emitted or absorbed only in definite amounts (quanta). This theory grew up around the concept of the discontinuity of energy. The wave theory believes that the radiant energy is composed in waves of the wave band of the radiation spectrum.

Quantum yield
The number of oxygen molecules released per light quanta absorbed or the quantum light energy required for reduction of one mole of carbon dioxide to the level of carbohydrates.

Quantum

The process of radiation takes place in discrete amounts called quanta (plural of quantum) and a quantum is the smallest indivisible unit of energy with no electric charge and very little mass. A quantum of electromagnetic radiation is 'photon' and the quantum of energy associated with nuclear forces is 'meson'.

Queststationary front

A front which moves very little since the last synoptic position.

R

'R' value

It is the percentage of the crop land actually cropped in a year or frequencies of cropping in a fallow cycle. R equal to crop year times 100. "R" values is synonymous to cropping index.

Rabi season

Season during which winter crops like wheat, barley, mustard, gram, linseed, etc. are cultivated and usually it extends from October to April.

Rad

The unit of radiation dose is 'rad' and this is defined as the total absorbed dose of 100 ergs of energy per gram of tissue in unit time.

Radar sonde

A system in which both meteorological and wind data may be obtained, as this is carried by radiosonde.

Radial velocity

The velocity that is parallel to the radar beam, towards or away from the given point.

Radian

The unit angle subtended at its center, in which the length of the arc is equal to its radius.

Radiance

Radiant flux per unit area per unit solid angle. Radiance has unit of watts per metre per metre per steradian. The radiance of a perfectly diffuse surface (Lambertian) is independent of the viewing angle.

Radiant energy

The energy that is emitted, transferred or received as radiation. To make this term more perfect and meaningful the word 'radiation' shall be extended to total electromagnetic radiation, i.e., wave band. The reason is that, the radiant energy is released in a continuous process, as long as the temperature of a body is more than absolute zero.

Radiant excitance

Same as 'Emittance'.

Radiant flux density

The radiant flux per unit area of a real or imaginary surface, usually oriented perpendicular to the rays source. The units are watts per metre or joules per metre per metre per second.

Radiant flux

The amount of electromagnetic radiant energy emitted, absorbed or transmitted per unit time. This is measured in watts or joules per second. Sometimes, the above three processes (emission, absorbance and transmission) are replaced with a term 'Flow', which implies the same meaning.

Radiant intensity

The radiant flux leaving a point on the source, per unit solid angle of space surrounding the point.

Radiation balance

Radiation balance of Earth's surface consists of two radiation streams of different spectral ranges. The short wave radiation, during day time, as long as Sun shines and the long wave radiation both by day and night. It is estimated that the Earth absorbs about 124 kilo langleys of solar radiation every year and radiates back 52 kilo langleys and the remaining 72 kilo langleys is the net balance (solar radiation) at the surface of the Earth.

Radiation chart

A chart or diagram readily available to calculate the heat or long wave radiation flux from the atmosphere. The water vapour, carbon dioxide, etc., which form major absorbing and emitting constituents for this radiation are taken into consideration for the distribution of temperature in preparing these charts.

Radiation ecology

The study of the effects of radioactivity on the organisms and their environment.

Radiation fog

When net radiation is negative, then the air in contact with the underlying ground surface is cooled. If sufficient moisture is available in the air, and cooling is also sufficient enough, then condensation occurs and fog is produced. This is called as radiation fog.

Radiation use efficiency (RUE)

This is the correlation between intercepted solar radiation by a crop canopy and biomass produced. The radiation use efficiency is equal to the weight of standing biomass divided by accumulation of intercepted radiant energy. This can be calculated either for the whole growing period or for defined phases of development of a crop.

Radiation

The process by which the energy is transferred, emitted or received without the use or necessity of a medium (space) in the form of electromagnetic waves or quanta. Sometimes, the emission of particles or waves by radioactive bodies is also referred to as radiation.

Radiational cooling

The cooling of the earth's surface which suffers a net loss of heat due to outgoing radiation.

Radioactive elements

Elements that radiate waves and particles either naturally or after having been made artificially radioactive.

Radiosonde balloon

A balloon larger than pilot balloon, which is used to carry a radiosonde aloft.

Radio sonde observation

An evaluation of air temperature, pressure and humidity of troposphere and low stratosphere through radio signals received from a balloon borne radiosonde.

Radio sonde

A balloon borne instrument package for the simultaneous measurement and transmission of meteorological data. It includes thermometer, barometer, hygrometer and radar reflector for the measurement of pressure, temperature and humidity. This is the standard sounding system of the international meteorological network for sensing the troposphere and low stratosphere during free ascent.

Radio theodolite

An instrument used to determine the direction from which radio waves approach a receiver.

Radiometer

An instrument used to measure the intensity of solar radiation. There are many organizations that manufacture different radiometers which are useful in measuring different components of solar radiation.

Radiometry

The measurement of properties of radiant energy.

Random sample

A sample for which every individual of the population has an equal chance of being chosen.

Rain atmometer

An instrument used to measure evaporation from plant canopy (particularly the effect of sunlight on evaporation).

Rain day

- A twenty four hour period, usually starting from 0730 IST during which a minimum amount of 0.2 millimetres or more of rain fall occurs.
- By international standards, a climatological period of twenty four hours from nine G.M.T. within which atleast 0.2 millimetres (0.01 inches) of precipitation is recorded.

Rain forest

The tropical and temperate forests which grows in the regions of heavy annual precipitation.

Rain gauge

An instrument used to measure total depth of water from precipitation (rainfall, melted snow, hail, drizzle, etc.). It is presumed that the water is distributed over an impervious horizontal surface and not subject to evaporation.

Rain out

A process in which radioactive dust and pollution particles that act as condensation nuclei in the precipitation formation are removed from the atmosphere and reach the ground. Same as 'Washout'. See 'Washout'.

Rain season

That period of the year in which monthly rainfall exceeds 10 per cent of the monthly potential evapotranspiration (In India May-June to September-October).

Rain shadow

A region in the leeward of mountains or ghats, over which moist air masses have to pass and subsequently rain falls. As compared to windward side the rainfall in the leeward side , (rain shadow area) is less, because the air masses become dry while moving.

Rain

The precipitation composed of liquid water drops larger than 0.5 millimetres in diameter (as distinct from drizzle). The drops may fall in straight paths but need not necessarily be in vertical direction.

Rainbow

A colour effect produced in the form of concentric arcs by the refraction and internal reflection of sunlight in minute droplets of rain, drizzle, or fog, in the atmosphere. The colours range from violet to red. This is seen only when the Sun is located behind the observer and the screen of water droplets on which rainbow formed is to the front.

Rainfall

The total depth of precipitation (rain, snow, dew, etc.,) measured in the rain gauge, at any particular location.

Rainfed dryland

Areas that depend on rainfall for crop production but are not flooded. The run-off and infiltration of rain water is so high that water does not accumulate on the land.

Rainfed farming

Growing of field crops entirely with rain water received during the crop season (rainfall usually > 750 mm) under humid and subhumid climate and the crops may face little or no moisture stress during their life cycle.

Rainfed wetland

Areas where rainfed rice is grown on puddled soil in fields bounded by dikes capable of ponding water to about 30 cm. Water depth seldom exceeds 30 cm. These areas receive no irrigation and non-rice crops may be grown before or after the wetland rice crop.

Randomization

An act of assigning the treatments to different plots by a random process.

Randomized block design (RBD)

In this design the land on which the trial is to be carried out is divided into as many blocks of the same size and shape as there are replications and each of the blocks into as many plots of the same size and shape as there are treatments.

Range of a function

Let 'f' be a function from a set A into a set B so that the domain of 'f' is the set A. Then the set of all the function values is called the range of 'f'.

Range of a sample

In calculus, the difference between the largest and the smallest values in a sample.

Range resolution

The ability of radar to distinguish between targets at different ranges on the same azimuth.

Range

The difference between minimum and maximum readings recorded from an instrument.

Rankine temperature scale

The absolute temperature scale on which the size of the degree is the same as on the 'Fahrenheit scale'. Zero degrees Rankine is $-459.67\ ^{\circ}F$.

Rational numbers

The set of fractions when enlarged by incorporating in it the set of negative fractions and zero, the entire set of these numbers is known as the set of rational numbers.

Ratoon cropping

The cultivation of an additional crop from the regrowth of stubbles of previous main crop after its harvest, thereby avoiding reseeding/replanting such as in sugarcane, sorghum, rice, fodder grasses etc.

Rawinsonde

A sophisticated version of the radiosonde for upper air observation. In this method of observation, like radiosonde, vertical profiles of temperature, pressure and humidity are measured. In addition to this, also measured are wind speed and direction.

Rawin target

A special type of radar target, usually a corner reflector, tied beneath a free balloon and designed to be an efficient reflector of radio energy.

Rawin

A method of winds aloft observation accomplished by tracking a balloon borne radar target or radiosonde either with a radar or a radio theodolite.

Ray model

To describe the behaviour of light, different models are developed and are being used. Of them this is one. This model is developed on the assumption that the light travels in straight lines from a source.

Ray

A narrow portion of a light beam of electromagnetic radiation.

Rayleigh scattering

This is the scattering of solar radiation (light) by particles of considerably smaller in size than the wavelength of the incident radiation. In the electromagnetic spectrum, for scattering of light in the interval of 300 to 700 nanometres, air molecules are responsible. Usually, if the circumference of the scattering particles is lesser than one tenth of the wavelength of the incident radiation, the scattering coefficient is inversely proportional to the fourth power of the wavelength of the incident radiation.

Raynold's analogy

The transfer coefficients of momentum, heat and matter are same, if turbulent transport among them is presumed to be exactly the same.

Real image

An image from which light actually comes to the eye.

Real number

A number, rational or irrational.

Recession curve

It is the relationship between the time at which water has completely disappeared from the soil surface and distance along the run in a surface irrigation system.

Recession farming (diara land farming)

It is a system in which crops are planted in flooded areas as the rainy season ends and water recedes. This system takes advantage of thoroughly saturated soil profile and also has the advantage of silt and nutrients left behind by flood water. This practice is prevalent in parts of Bihar to Assam.

Recession phase (irrigation)

Portion of total irrigation time between the beginning of recession at the upper end and the disappearance of the water from the field surface. Time elapsed during this phase is known as recession time.

Recharge

Replenishment of ground water storage from precipitation, infiltration from surface streams and other sources.

Reclamation (land)

The operation or process of changing the condition or characteristics of land which can not be utilized to its full potential otherwise viz., irrigation of land, drainage of swamp or waterlogged land, gypsum application to alkali soils, liming of acid soils etc.

Reconnaissance code (RECCO)

A detailed weather observation or investigation from an aircraft in flight.

Red soils

They are formed mainly from granite rocks. They are highly variable in depth and show a well developed profile character. They are more coarse grained than the black soils. The percentage of clay increases markedly from the top soil to the sub-surface soil. The sub-soil at times may be compacted and may not be very permeable to water. The soils grade from the poor, thin gravelly and light coloured soils of the uplands to the more fertile, deep dark varieties, poor in nitrogen, phosphorus and humus but fairly rich in potassium. Murrum is the gravelly disintegrated rock found beneath the red soils.

Reduced sample

A representative part of the aggregate sample obtained by a process of reduction in such a manner that the mass approximates to that of the final (laboratory) sample.

Reduced tillage

A tillage system in which the primary tillage operation is performed in conjuction with plating procedure in order to reduce or eliminate secondary tillage operations. A type of soil conservation tillage in rain fed or irrigated areas.

Reducing and non-reducing sugars

Sugars possessing a free aldehyde or ketone group are reducing sugars, such as maltose. When the above groups are locked up in a glycosidic linkage in the formation of complex sugars, the resulting sugars are said to be non-reducing, for example, sucrose.

Reflectance

The ratio of a monochromatic beam of electromagnetic radiation reflected by a body to that incident upon it and usually expressed in percentage.

Reflected solar radiation

The solar radiation directed upwards after reflection by the Earth's surface and atmosphere.

Reflection coefficient

The fraction of incident radiation intensity reflected by a surface. Sometimes, albedo of ground surface is also referred to by this term.

Reflection

Bending back of waves.

Reflectometer

A Pyranometer is called a reflectometer when it is used to measure reflected solar radiation by making it facing downwards. Usually, to record 'Albedo' values of a surface, a pyranometer is kept inverted.

Refraction index of a medium

The ratio of velocity of light in vacuum to the velocity of light in the medium.

Refraction

The bending of any wave as it passes from one medium to another, in which the speed of travel is different.

Refractive index

The ratio of the velocity of light in certain medium to its velocity in air under the same conditions.

Regeneration

Sprouting of stubbles following cutting and/or grazing, Synonymous to regrowth or fresh growth.

Regional climatology

A descriptive climatology concerned with the identification of important climatic characteristics and which analyses the interaction of the weather and climatic elements upon the life, health and economics of people and area.

Regression coefficient

- A coefficient that gives the rate of change in one variable (dependent variable) per unit rate of change in another (independent variable).
- A numerical measure of the rate of change of the dependent on the independent variables.

Regression equation

An equation expressing the relationship between the specified value of one variable and the mean of all corresponding values of another variable dependent on the first.

Regression

- In statistics, the derivation of a relationship between a dependent variable and one or more independent variables.
- Relationship between specified values of one variable and the means of all corresponding values of another variable dependent on the first.
- Is a quantitative measure of relationship between two variables which helps to predict the value of one variable (dependent) from the known value of the other (independent) variable.

Regular reflection

The reflection of undisturbed electromagnetic waves which may result in the formation of an image of the source of light.

Relation between pressure and kinetic energy

The pressure of a gas is equal to 2/3 of the translational kinetic energy of the molecules in a unit volume.

Relative crop intensity index

It determine the amount of area-time allocated to one crop or a group of crops relative to the area-time actually used in production of all crops.

Relative crowding coefficient (k)

It is a measure of the relative dominance of one component crop over the other in an intercropping or mixed cropping system. The coefficient (k) is determined separately for each component crops e.g. for crop 'a' in association with b in 50:50 mixture the coefficient,

$$k_{ab} = \frac{\text{Mixture yield of a}}{\text{Pure stand yield of a - Mixture yield of 'a'}}$$

If k > 1 or equal to 1 or < 1, then there is advantage or no difference or disadvantage.

Relative density (RD)

The population of a particular species expressed as a percentage of all the species in a habitat.

Relative humidity

The dimensionless ratio (usually in percentage) between the actual amount of water vapour present in the air and the amount of water vapour needed for saturation under the same temperature and pressure. A measure of degree of saturation of air.

Relative spread index

It represents an area of the crop expressed as percentage of the total cultivated area in the zone divided by the area of the crop expressed as percentage of the total cultivated area in the country, times hundred.

$$RSI = \frac{\text{Cultivable area under a crop in the district}}{\text{Total cultivable area under that crop in country}} \times 100$$

Relative yield index

This implies that the mean of plant relative yield (ratio of the per plant yields in a mixture) to that in monoculture will have a value close to unity. In terms of relative yields, the value is a total, which has been called the relative yield total.

$$RYT = \frac{Y_{ab}}{Y_{aa}} + \frac{Y_{ba}}{Y_{bb}}$$

where

Y_{aa} = 'a' yield under sole crop,

Y_{bb} = 'b' yield under sole crop

Y_{ab} = 'a' component yield in association with 'b'

Y_{ba} = 'b' component yield in association with 'a'

Relay cropping

Seeding/planting of succeeding crop after flowering and before the harvest of the standing crop, it is analogous to a relay race where one crop hands over that baton (land to the next crop in quick succession, e.g. sowing toria in standing crop of maize).

Relief

The configeration of the land or the vertical variation in terrain including such features as direction and degree of slope, sharpness of change in slope direction and variation of these features.

Remote sensing

Direction and or measurement of atmospheric quantities from a distance by active (eg. Radar) or passive (eg radiometry) instrumentation, as opposed to 'in situ' measurement.

Residence time

The time spent by a particle from one state to other state in a crop canopy.

Residual effect of manure (fertilizer)

This refers to the residual beneficial effect of application of organic manure/fertilizers on the succeeding crops due to unutilized plant nutrients left over by the preceding crop

Residue (pesticides)

That quantity of a substance especially of active pesticide, remaining in soil or plant. Now-a days there is growing concern about the alarming levels of chemical residues in food stuffs that cause ailments to humans to live stock.

Resilience (crop)

The resilience of a crop is the extent of its ability to survive a perturbation or stress with minimum effect on yield. It is, thus, critical component of the crop yield stability. In a sole crop, recovery after a shock may involve compensation within and between plants of similar genotypes. In an intercrop, compensation is often also between components.

Resistance thermometer

Electrical thermometer whose principle is based on the fact that the electrical resistance of a conductor varies with temperature, normally increasing with rise in temperature. In this thermometer, the temperature is deduced from the measurement of the resistance of a metal, usually platinum or nickel in the form of a wire.

Resistance

The degree to which a species of plant or other organism tolerates a toxic substance or moisture stress, or diseases.

Resolution

The ability of a radar to read two distinct targets separately.

Respiration

The process by which energy is acquired in a living plant organism, by the breakdown of complex organic molecules, with the release of waste carbon dioxide and water.

Response curves

A close functional relation between the input x (nutrients) and output Y (production) can be expressed in some algebric equation in the form $Y = f(x)$

Restoration grazing

Intensive system of management whereby grazing is deferred on various parts of the range during succeeding years, allowing the deferred part complete rest for one year; two or more units are required.

Result demonstration

A type of demonstration in which the results of proven technology are shown in comparision to existing practices.

Resultant force

A force resulted by the addition of other forces acting on a body.

Return period

The anticipated or probable period of time between the repetition of two extreme events.

Reversal of the monsoon

The arrival of winter season in place of summer season or summer season in place of winter season.

Reversible

A process which can be reversed with a small change in the system by passing through its equilibrium.

Reversing layer

The transparent and deep solar atmosphere outside the photosphere of the Sun, which occupies several hundreds of kilometers.

Reynold's numbers

The square of the Brunt Vaisal frequency divided by the square of the vertical shear of the horizontal wind.

Rice glazing

Polished rice treated with talc (and in some cases with glucose).

Rice milling

Involves the removal of hulls and bran from paddy rice. Milling ranges from hand pounding to modern milling equipments. It includes various steps viz, dehulling, polishing and pearling (scouring).

Richardson's number

In flow regimes, a dimensionless number (a ratio) associated with the shearing flows of a stratified fluid. In the calculation of wind velocity and associated profiles this is a measure of dynamic stability.

Ridge (of high pressure)

An elongated area (zone) of relatively high pressure, usually associated with an area of maximum anticyclonic curvature of the wind flow.

Ridge former (ridger)

An impliment used to form ridges and furrows or channels. It is important implement for planting sugar cane, Maize, cotton, castor etc.

Ridge planting

Planting on ridges

Rill erosion

Erosion producing small channels that can be destroyed completely by tillage.

Rime

Freezed fog on any objects. The granular ice particles grown on windward side of exposed objects from supercooled or freezed fog or cloud and subsequent freezing. This is built out directly against wind.

Rip Current

It is formed by a strong surface water movement, of a short duration that flows seaward from the shore.

Ripening stage

Period between completion of grain formation to maturity.

Rocket sonde (Meteorological rocket)

A rocket designed primarily for routing upper air observations in the lower 250,000 feet of the atmosphere, especially that portion which is inaccessible to balloons (above 100,000 feet).

Rogue

An off-type plant in a crop field. Removing of off types is called 'roguing; it is done once or twice in seed production programme to maintain purity.

Roll cloud

A low-level, horizontal, and tube-shaped cloud which is associated with a thunderstorm.

Root crop

Crop plants whose surplus or reserve foods are stored primarily in enlarged roots e.g. sweet potato.

Root nodules

Refers to small swellings on roots of leguminous plants produced as a result of infection by nitrogen-fixing bacteria (Rhizobia). If cut and seen they appear pinkish that indicates their functional activity, other wise it appears black.

Rossby number

The dimensionless ratio of relative accelerations and coriolis accelerations typical of a particular regime of synoptic scale flow.

Rossyby wave

When the temperature gradient across a rotating fluid (such as in the annulus experiment) reaches a critical value, the previously symmetric flow breaks down into a wave like flow. The waves being termed Rossby waves after the Sweedish meteorologist who first recognized the importance of transient mid-latitude disturbances to the general circulation of the atmosphere. A wave of global length scale (of the order of thousands of kilometres) propagating by virtue of the planetary vorticity (coriolis parameter) with latitude.

Rotary tillage

Tillage operation employing rotary action to cut, break and mix soil.

Rotation pasture

A pasture used for a few seasons and then ploughed for others crops.

Rotational grazing
Dividing a large pasture into several small paddocks and allowing the cattle to graze each, periodically perhaps every four weeks.

Rotational velocity
The velocity of rotation on an axis.

Rotor cloud
An altocumulus cloud in the lee of a mountain in which the air rotates around a horizontal axis, creating turbulence.

Roughage
Plant materials (feedstuffs) that are relatively high in crude fibre and low in digestible nutrients such as straw and stover.

Roughness length
This is a constant of integration in logarithmic velocity profile appropriate to fully rough flow near a surface. In wind flows, this is expressed as a non-dimensional ratio with zero plane displacement. This quantity is also called as 'roughness coefficient' or 'roughness parameter'. Also see 'Roughness parameter'.

Roughness parameter
In aerodynamics this is the measure of the roughness of a surface over which wind is moving. Same is the case in fluid flow regimes. Also see 'Roughness length'.

Row intercropping
Growing two or more crops simultaneously where one or more crops are planted in rows.

Run of wind
The distance traveled by air in its movement which is recorded in an anemometer for a unit time. For all meteorological purposes, it is calculated for an hour or day. In many micrometeorological measurement related to crop studies, it is recorded for 3 minutes.

Run-off
The portion of precipitation, particularly the water from rain or melting snow, which flows over the surface (an area or land) and ultimately discharged through stream channels.

S

Saffir-Simpson damage -potential Scale

A measure of hurricane intensity on a scale of 1 to 5, in which the scale categories potential damage based on barometric pressure, wind speed, etc.

Saline alkali soil

A soil containing a combination of soluble salts and exchangeable sodium sufficient to interfere with the growth of most crop plants. The electrical conductivity and exchangeable sodium of the saturation extract are more than 4 milli mho per centimeter at $25^{\circ}c$ and more than 15% respectively. The pH is usually 8.5 or less in the saturated soil paste.

Saline soil

A soil containing excess of soluble salts to impair soil and crop productivity and has electrical conductivity value > 4 mhos/cm

Salinization

The process of accumulation of soluble salts in a soil usually chlorides and sulphates.

Salt index

Concerning fertilizer salts and compound fertilizers; an index of the extent to which a given amount of fertilizer increases the osmotic pressure of soil solution.

Salt

The product other than water of the reaction of a base with an acid.

Saltation

Soil particle movement in water or wind where particles skip or bounce along the stream bed or soil surface.

Sample

A relatively small group of individuals taken from a very large population about which we seek information.

Sampling error

Deviation of a sample value from the true value owing to the limited size of sample.

Sampling unit

A defined quantity of material having a boundary which may be physical for example a container, or hypothetical, for example a particular time or time interval in the case of a flow of material.

Sampling

The procedure of selection of a sample from population is known is sampling.

Sand

Small rock or mineral fragments having diameters ranging from 0.05 to 2.0 mm.

Sandstorm

A strong and violent wind carrying sand particles through the air, that reduces the visibility to greater levels.

Sandy clay loam

This class of soil contains 20-35% clay, less than 28% silt and 45% or more of sand.

Sandy clay

Soil of this textural class contains 35% or more of clay and 45% or more of sand.

Sandy loam

Soil of the sandy loam class of textures which has 50% sand and less than 20% clay.

Sandy soil

A broad term for soils of the sand and loamy sand classes where more than 70% sand and less than 15% clay are present.

Saprophytic

Obtaining nourishment from non-living biological materials.

Satellite Images

Images taken by a satellite that reveal information of interest.

Satellite

An object that orbits a celestial body. However, it refers to the manufactured objects that orbit the earth, either in geostationary or a polar manner.

Saturated adiabat

Any statistically valid thermodynamic diagram on a graph related to adiabatic change under saturated conditions of an air parcel.

Saturated adiabatic lapse rate

In a column of saturated air the rate of decrease of temperature, when lifted adiabatically. Here, the latent heat of condensation or sublimation partially compensates for cooling due to expansion. This is largely variable, and associated with both adiabatic ascent or descent of a saturated air parcel.

Saturated air

A column of moist air in a state of equilibrium with plane surface of either pure water or ice at the same temperature and pressure. In other words, air is said to be saturated when its vapour pressure is equal to saturated vapour pressure and relative humidity one hundred percent.

Saturated soil

A soil which has its interspaces or void spaces completely filled with water to the point where run off occurs.

Saturation deficit

This is the difference between the saturation vapour pressure and the actual vapour pressure of a given volume of air at the existing temperature. So, this is the amount of water vapour required to bring a given volume of nonsaturated air to the point of saturation at a given temperature and pressure. Also called as vapour pressure deficit. Also see 'Vapour pressure deficit'.

Saturation index

An estimate of carbonate precipitation from irrigation water as a function of the degree of calcium cabonate saturation of the soil saturation.

Saturation light intensity

Light intensity of photosynthesis are directly proportional to each other only upto certain level. If this physiological phenomenon the saturation light intensity is that level after which any additional light intensity will not produce an increase in photosynthesis. The level at which the photosynthetic rate becomes independent of light.

Saturation of atmosphere

The state of the atmosphere, at which the maximum possible amount of water vapour is held at a given temperature.

Saturation point

The point at which the density of the water vapour is maximum above which it no longer increases.

Saturation vapour pressure

In micrometeorology, this term is often used to denote the pressure exerted by a parcel of saturated vapour and this is a function of temperature. In general meteorology, this term refers to the partial pressure of water vapour in equilibrium with a plane surface of water, ice, etc.

Saturation

If a column of air holds the uppermost limit of water vapour that it can held, then the air is said to be saturated. Perhaps, the relative humidity of one hundred percent is saturation. A condition that should be obeyed by an air parcel in this situation is that the same parcel should be in equilibrium with the surface of a pure water or ice.

Scale height

The vertical elevation at which a particular atmospheric property falls to a certain proportion as compared to its surface value. Here, the atmospheric property may be either air density or pressure or any other parameter.

Scalar

A measurable quantity with no direction but possessing its own magnitude.

Scarification (seed)

Scratching of hard seed coat to remove the mechanical barrier to obtain uniform germination with better penetration of moisture in to the seed.

Scattered radiation

This is also a part of incoming solar radiation. When solar radiation passes through the atmosphere a part of it is scattered by water vapour, air and other gas molecules. Some more portion is diffusively scattered and reflected by clouds in the troposphere. This total portion which is scattered by all these particles is known as scattered radiation.

Scattering

- The process of chaotic deflection of electromagnetic radiation into different direction by a particle or scattering surface. It is a fact that the atmosphere is composed of air, water vapour, different gases, ions, dust particles, smoke, etc. These particles obstruct the movement of energy waves of shorter diametre. So, naturally the same are deflected into various directions. An important aspect involved is that the angle at which the rays are hitting the molecules which again depends upon the exposure of the Earth to the Sun.

- The deflection of light by particles. This effect is most pronounced for high frequencies such as blue light, giving the sky a blue hue.
- This is a process by which molecules of a medium and small particles suspended in the medium diffuse a portion of incident radiation in all directions.

Scatterometer

An instrument used to measure the amount of scattered portion of a radiation beam or a scattering surface.

Scientific name

The binomial latin name of a species consisting of the generic name, a latin of latinized noun, followed by the species epithet a latin or latinized adjective.

Screen

The single or double Stevenson screen is popularly known as 'screen' in the field of meteorology.

Scud

Low fragments of clouds, which are unattached below a layer of higher clouds.

Sea breeze

In coastal areas, during day time, land surface gets heated up quickly and it is hotter than sea surface. So, the air over land rises because of decrease in pressure and increase in air over the sea surface and relatively hot air over the land surface. This situation cause on shore flow of air during day which is known as sea breeze.

Sea Fog

A type of advection fog which forms in warm moist air cooled to saturation as the air moves across cold sea water, or a large water body.

Sea Ice

Ice formed due to freezing of sea water.

Sea level pressure

Atmospheric pressure at mean sea level deduced from the observed station pressure.

Sea level

The height of the sea surface at any time.

Sea mile

A unit of length which is equivalent to 1,000 fathoms (6,000 feet).

Sea smoke

This is a type of advection fog. It is formed primarily over water when cold air passes across warmer waters.

Sea spray

The drops of sea water blown from the top of a wave, which is also called as salt spray.

Season

A division of the year according to some regularly recurring astronomic or climate phenomena.

Seasonal consumptive use

The total amount of water consumed in evaporation and transpiration by a crop during the entire growing season expressed in depth or volume of water per hectare.

Seasonal drought

A kind of drought which occurs during distinct annual periods of dry weather.

Secondary radiation standards

The instruments used in the radiation measurement are often compared with the standard ones. The ones that are maintained at the regional and national radiation centers are regularly compared with the ones at internationally designated centers and the latter are called primary radiation standards and the former are secondary radiation standards. The process is not only customary but also mandatory, as there is need to maintain accuracy and precision of the recordings.

Secondary tillage

The tillage operations following primary tillage to create a good seed bed for proper seeding/planting.

Seed bed

A well prepared land for sowing/planting.

Seed certification

A means to maintain and make available to the public, sources of high quality seeds and propagating materials of superior varieties so grown and distributed as to insure genetic identity. This is done by means of inspections of fields and seeds and by regulations for checking on the production, harvesting and cleaning of each lot of seed.

Seed control

Seed law enforcement which applies to all seeds sold commercially. Correct labeling with respect to the kind of seed, germination, purity and other quality factors are ensured.

Seed dressing

To chemically treat seeds before sowing to control any disease or pest attack in growing crops, particularly cereals.

Seed drill

A machine to place seeds at uniform rate at selected depth and row spacing.

Seed rate

The quantity of seed required for sowing on an unit area of land.

Seed treatment

The process of application of a thin coat of insecticide or fungicide over seeds or similar applications to prevent pest infestation.

Seed-drill

Sexually or vegetatively propogated planting materials which are used for seeding and planting and as such should be free from (pests and diseases) any infection and should give a good crop stand by good seeding.

Seed-cum-fertilizer drill

An implement which drills seed and fertilizer simultaneously and uniformly in the same or different rows.

Seede

A device for placing seeds on or in the soil.

Seeding of clouds

Same as 'Cloud seeding'.

Seedling stage

It includes the period from emergence till just before the appearance of first tiller/branch.

Seedling

The juvenile stage of a plant grown from seed. Usually indicates plants which have up to and including about 4 true leaves.

Seepage

The process by which water passes through the soil laterally from one field to another. In rice fields the seepage loss is one of major aspects that shall be considered in water management.

Selection

In a population the preservation of certain individuals that have desirable characteristics.

Selective absorption

The way in which the selective reflection occurs is the way in which the selective absorption also occurs. When the solar radiation passes through the crop canopy certain wavelengths are absorbed by air inside the canopy and also by water vapour, in addition to plant parts. Each of these component absorbs the wavelength as per their dimensions. So, a particular wavelength absorbed by a water vapour molecule may or may not be absorbed by a crop component. This is specific absorption. However, irrespective of the wavelength absorbed, finally the radiant energy is converted into heat energy on absorption. Also see 'Selective reflection'.

Selective reflection

Of the seven colours of the visible part of the spectrum, if the reflection of one colour is greater than the remaining ones, then the reflection is said to be selective.

Self-mulching soil

A soil in which the surface layer becomes so well aggregated that it does not crust and seal under the impact of rain but instead serves as a surface mulch upon drying.

Semi arid zone

A zone in which rainfall is insufficient to cultivate crops in some years. Even during a year the evapotranspiration is more than moisture available for plants. The well defined summer and rainy seasons are common in some areas of the world.

Semiorganic fertilizer

Product in which declared nutrients are of both organic origin obtained by mixing and/ or chemical combination of organic and inorganic (mineral) fertilizers.

Semipermanent pressure systems

A stable, stationary pressure-and-wind system where the pressure is predominately high or low with the changing season.

Sensible heat advection

The process in which warm dry air passing over a field supplies energy for transpiration.

Sensible heat flux

Same as enthalpy and this is the product of heat capacity times the Kelvin temperature, at constant pressure for a perfect gas. This is used in meteorology in contrast to latent heat. In crop canopies the heat energy utilized in raising the temperature of air is referred to as sensible heat.

Sensor

- A part of an instrument used to convert an input signal into a quantity that can be measured by another component or part of the instrument.
- The component of an instrument that converts an input signal into a quantity that is measured by another part of the instrument.

Severe local storm

A storm with relatively strong magnitude as compared to a normal storm. Also see 'storm'.

Severe thunderstorm

A thunderstorm with winds measuring 50 knots (58 mph) or greater, which produce torrential rain, frequent lightning, hails etc.

Severe weather

The destructive weather events, like blizzards, intense thunderstorms, tornadoes etc.

Sewage farm

A farm which takes sewage, usually as sludge, from settlement tanks; the sewage is used as manure and the effluent is drawn off to irrigate the land. A farm with a sandy soil is preferred for sewage disposal since sand is porous and acts as a filtering material.

Shade plant

A crop plant that can grow and exhibit its potential in the shade.

Shadow

The part which is devoid of visible part of the spectrum because of the obstruction caused by a matter.

Shear Line

A line of maximum horizontal wind shear.

Shear

The rate of change over a short duration. In wind shear, it refers to the frequent change in wind speed within a short distance.

Sheet erosion

The removal of a fairly uniform layer of soil or material from the land surface by the action of rainfall and run off water.

Shelling

Removal of grains from cobs or kernels from pods capsules or husk from paddy grain.

Shelter factor

The ratio of the actual drag coefficient observed to the coefficient measured (or estimated) for the same element in isolation.

Shelter

A resting place or protective coverage provided to a crop to survive from unfavourable conditions of the physical environment.

Shelterbelt

Shelterbelts are the artificial barriers that protect the crop plants by altering the pattern of mean wind velocity, turbulence, radiation and energy balance of crop plants both in the lee and for a short distance windward. The barriers may be formed by another crop in a main crop field, dried straw, etc.

Sherwood number

The ratio of actual mass transfer to the rate of transfer that would occur if the same concentration differences were established across a layer of still air of thickness equal to zeroplane displacement.

Shifting cultivation

The practice of cultivating clearings scattered in the reservoir of natural vegetation (forest or grass-woodland) and of abandoing them as soon as the soil is exhausted; and this includes the practice of shifting homesteads in order to follow the cultivator's search for new fertile land. It is otherwise called as Jhum cultivation or slash and burn system.

Short day plant

A plant in which flowering can be induced or enchanced by short days, usually of less than twelve hours of daylight. These are also known as long night plants, e.g. rice, soybean.

Short range weather forecast

A type of weather forecast for agriculture, the validity of which is generally considered to be extended from a day to two days.

Short wave radiation

This is the part of solar spectrum having wavelengths less than 4 microns. This term is often used to radiation phenomena in the atmosphere as well as in crop canopies to distinguish it from solar, diffuse, sky and long wave radiations, etc. The terrestrial radiation is often referred to as long wave radiation to distinguish it from short wave radiation of the much hotter Sun.

Shower

Precipitation from a convective cloud which is characterized by its short duration and light intensity.

Side dressing

The application of commercial fertilizer along the side of a row or around a plant.

Significance level (stastics)

A probability value that it is considered so rare in the sampling distribution specified under the null hypothesis that one is willing to assert the operation of nonchance factors. Common significance levels are 0.05 and 0.01

Silage

Animal feed resulting from the storage and fermentation of green or wet crops under anaerobic conditions.

Silt loam

Soil material having 30 per cent or more silt and 12 to 27 per cent of clay or 50 –80 per cent of silt and less than 12 per cent of clay.

Silt

Small mineral soil particles of a diameter of 0.002 to 0.05 mm.

Silting

The deposition of water-borne sediments in lakes, reservoirs. Stream channels or over flow areas.

Siltyclay loam

Soil having 27-40 per cent of clay and less than 20 per cent of sand.

Silviculture

Is the art and science of cultivating forest crops as distinguished from silvics which deals with the study of the life history and general characteristic of forest trees.

Simoon

An intensely hot and dry wind in the deserts which usually carry much of sand with it.

Simple correlation

It measures how two variables (for example yield and spikelet number/plant) are associated in a sample. The correlation coefficient can be regarded as a measure of the intensify of the linear association and the value ranges from 0 to 1.

Simulation models

The art of building mathematical models and the study of their properties in reference to those of systems is known as simulation. When crop models are built on simulation they are called crop simulation models. When these are driven by daily weather variables they enable the quantitative description of the dynamic crop productive system.

Single grain soil

A structureless soil in which each particle exists separately as in sand dunes.

Sink strength

The sink size times its activity. In calculations the number of panicles, grains or fruits are taken as sink size and the relative growth rate is considered as its activities.

Sink

When the process of photosynthesis is over, the end products are stored at a local region in the plant as mass with energy. This region is called as sink.

Skew T-log P diagram

A thermodynamic diagram, using the temperature and the logarithm of pressure as coordinates to evaluate and forecast air parcel properties like the convective condensation level, the lifting condensation level, the level of free convection (LFC) etc.

Skin friction

The force that air exerts on a surface in the direction of the flow is a direct consequence of momentum transfer through the boundary layer and is known as skin friction.

Skip row planting

When a line is left unsown in regular sowing.

Skip row irrigation

When a row is left unirrigated, during field irrigation, usually followed in widely spaced crops like sugarcane, cotton etc., to improve irrigation efficiency.

Sky cover

The amount of the celestial dime that is hidden by clouds.

Sky light coloured

The sky is said to be light coloured when the existing cover of the cloud is equal to one or two oktas.

Sky radiation

The downward reflected component of thermal radiation after scattering in the atmosphere. The term is used as an alternate for scattered radiation.

Sky

In a locality or station, at any given period, the condition of the huge vacuum above the Earth with reference to the amount, height, etc., of the existing clouds.

Sleet

In different countries this is used in different meanings. Basically, this is a form of precipitation falling in small particles of clear ice. These particles may have formed either due to melting of hail or snow as it descends or when rain drops are frozen as they pass through a layer of cold air.

Slow release fertilizer

Fertilizer whose nutrients are present as a chemical compound or in a physical state such that their availability to plants is spread over a period of time. Example neem coated urea.

Sludge

The semi-solid part of sewage water that has been sedimented or acted upon by bacteria.

Slurry

Semi-liquid effluent from livestock, consisting of urine and feaces, possibly diluted with water.

Slush

Snow or ice on the ground that has been reduced to a softy watery mixture.

Smog

Basically, the term is safely used to refer to a situation in the lower troposphere when smoke and fog are mixed, which results in poor visibility. Of late, this term is used for any form of air pollution.

Smoke

Small particles produced by combustion that are suspended in the air.

Snow

A form of precipitation formed by sublimation of water vapour at below freezing point temperatures. The snow fall occur as hexagon feathery crystals of ice, either singly or in conglomerate flakes.

Snowfield

An area or mass of snow that remains over the field. A deposit of soil material accumulated in a mass of snow following melting of snow is called snowflush.

Soaking

The part of presowing seed treatment or parboiling where paddy is soaked in water in order to raise its moisture content.

Social forestry

It is a programme of forestry development and conservation under various agro-climatic and soil condtions through (a) mixed plantation on waste lands and panchayat lands and (b) reafforestation of degraded forests and raising shelter belts.

Sod culture

A system of soil management wherein the plants are grown in permanent grass without tillage.

Sod seeding

Mechanically placing seed, usually legumes, small grains, directly into a grass sod.

Sodar

A signal scattered back from a situation of temperature variation in the atmosphere.

Soil (mature)

A soil which has reached the full development to be expected under existing weathering and biological process.

Soil adhesion

The sticking of soil to tillage tools or wheels.

Soil aggregate

A naturally occurring cluster or group of soil particles in which the forces holding the particles together are much stronger than the forces between adjacent aggregates.

Soil air

Below the surface of the Earth, the air and other gases that are present in the soil atmosphere and which are essential for plants life and activities of micro organisms.

Soil amendment

Any substance that is added to the soil for the purpose of improving its physical or chemical character, enhancing soil productivity or promoting the growth of crops but excluding commercial fertilizers and organic manure.

Soil auger

A tool for boring into the soil and withdrawing a small sample of soil for laboratory analysis.

Soil compaction

The colloidal fraction of the soil primarily made up of inorganic material (clay) with varying amounts of organic colloids (humus).

Soil conditioners

These are chemicals which are added to maintain physical condition of the soil e.g. polyvinylites, polyacrylates, cellulose gums, lignin derivatives and silicates.

Soil conservation

A combination of all management and land use methods which safeguard the soil against depletion or deterioration by natural or by man-induced factors.

Soil consistency

It is the capacity of the soil to resist manifestation of the physical forces of cohesion and adhesion, acting within the soil at various moisture contents.

Soil cultivation

Tillage operations performed to create soil conditions conducive to improve aeration, infiltration, compaction and moisture conservation as well as to control weeds.

Soil cutting

Soil separation from the soil mass by slicing action.

Soil degradation

Is a process of changing a soil from one type to another more highly leached one; particularly bringing about replacement of sodium by hydrogen by leaching a saline or alkali-soil.

Soil erodibility factor

The soil erodibility factor indicates the inherent erodibility of a soil. It gives an indication of the soil loss from a unit plot 22 mm long with a 9% slope and continuous fallow culture.

Soil evaporimeter

An instrument used to measure the water evaporated, at any given period of time from the soil.

Soil failure

The alternation or destruction of a soil structure caused by mechanical forces, such as shear, compression and impact.

Soil fertility

Soil fertility refers to inherent capacity of a soil to supply nutrients to plants in adequate amount and in a suitable proportion for higher productivity. All productive soils are fertile, but all fertile soils may not be productive.

Soil heaving

The swelling of soil resulting from natural forces, such as freezing.

Soil management

The sum total of all tillage operations, cropping practices, fertilizers, lime and other treatments, conducted on or applied to a soil for the production of plants.

Soil moisture deficit

The amount of water needed to return the soil moisture status to the field capacity.

Soil moisture stress

It is the sum of soil moisture tension and the osmotic pressure of the soil solution.

Soil moisture tension

It is the force with which the water is held by soil against gravity. It is generally expressed in atmospheres or in mm of mercury.

Soil moisture

In micrometerological studies of crop plants, this term is used strictly to refer to the humidity contained in root zone of the plants. In the study of other soil physical properties, etc., this term refers to the moisture contained in the soil above the water table including the water vapour present in the soil pore.

Soil morphology

A systematic study and description of the characteristics of the soil profile in the field.

Soil organic matter

The organic fraction of the soil that includes plant and animal residues of various stages of decomposition, cells and tissues of soil organisms, and substances synthesized by the soil population.

Soil persistence (herbicide)

Refers to the length of time that a herbicide application on or in soil remains effective.

Soil productivity

Is the present capacity of a soil to produce crop yield under a defined set of management practices. It is measured in terms of the yield in relation to the input of production factors.

Soil profile

A vertical section of the soil from the surface through all its horizons into the parent material.

Soil reaction

The acid or alkaline reaction of a soil as measured by one or number of methods expressed in terms of pH units, generally in the range of 4.5 to 9.5.

Soil salinity

The amount of soluble salts in a soil, expressed in terms of percentage, parts per million (ppm), or other convenient ratios like, mhos/cm or decisimens m^{-1}.

Soil temperature
The temperature recorded at required depths in the soil.

Soil texture
Is an expression or relative proportion of primary soil particles viz., sand, silt and caly in soils. A soil is described as a coarse, medium or fine textured depending on the predominant particle size.

Soil type
A sub division of a soil series based mainly upon the texture of the upper layers.

Soil water potential (capillary)
The amount of energy that must be expended per unit quantity of pure water in order to transport reversibly and isothermally an infinitesimal quantity of water, identical in composition to the soil water, from a pool at the elevation and the external gas pressure at the point under consideration to the soil water.

Soil water potential (gas pressure)
This is a potential component to be considered only when external gas pressure differs from atmospheric pressures as in a pressure membrance apparatus.

Soil water potential (gravitational)
The amount of energy that must be expended per unit quantity of pure water in order to transport reversibly and isothermally an infinitesimal quantity of water identical in composition to the soil water, from a pool at a specified elevation and atmospheric pressure, to a similar pool at the elevation of the point under consideration.

Soil water potential (osmotic)
The amount of energy that must be expended per unit quantity of pure water in order to transport reversibly and isothermally an infinitesimal quantity of water from a pool of pure water, at a specified elevation and at atmospheric pressure, to a pool of water identical in composition to the soil water (at the point under consideration), but in all other respects being identical to the reference pool.

Soil water potential (total)

The amount of energy that must be expended per unit quantity of pure water in order to transport reversibly and isothermally an infinitesimal quantity of water from a pool of pure water at a specified elevation and at atmospheric pressure to the soil water (at the point under consideration), consists of osmotic, gravitational and capillary potential.

Soil

- A collection of natural bodies developed in the unconsolidated mineral and organic material on the immediate surface of the earth that serves as a natural medium for the growth of plants and has properties due to the effects of climate and living matter acting upon parent material, as conditioned by topography, over a period of time.
- The agrometeorlogists consider soil as a layer of loose material which covers the cropped or bare land on the surface of the Earth.
- A body formed from a mixture of broken and withered minerals and decaying organic matter which supplies nutrients and water for plant growth.

Soilage

Forage cut green and fed to livestock while in the fresh condition are known as soiling crops or as soilage. Example berseem.

Soilmoisture characteristic curve (moisture retention curve)

A graph showing the relationship between the amount of water remaining in the soil at equilibrium as a function of matric suction. Otherwise they are called as soil moisture release curves and differs based on type of the soil.

Solar constant

The energy falling in one minute on a surface of one centimetre per centimetre at the outer boundary of the atmosphere held normal to the sunlight, at the mean distance of the Earth from the Sun. It is estimated to be 1.94 to 2.0 calories per centimetre per centimetre per minute.

Solar day

The time taken for one complete rotation of the earth in relation to the sun.

Solar eclipse

The eclipse that occurs when the moon is in a direct line between the sun and the earth, casting some of the earth's surface in its shadow.

Solar elevation

The angular elevation of the Sun above the horizon

Solar farm

A suggested power-utility, based in a desert and converting a considerable area, where large amounts of electrical energy would be generated from solar energy by capturing it by photovoltaic devices.

Solar noon

The moment when the Sun crosses the meridian of observation.

Solar spectrum

A very broad part of the electromagnetic spectrum occupied by wavelengths from which the radiant energy received from the Sun is spread over.

Solar

In meteorology and micrometeorology of crop plants any activity associated with 'Sun' is given a prefix with this term.

Solarimeter

An instrument used to measure the radiation received from the Sun.

Solarization

In plant physiological terms the inhibiting effect of extremely high light intensities of photosynthesis. In agricultural practices, this term is often used in post harvest sun-drying of produces of different crops.

Sole crops

A crop grown in pure stand at optimum population (spacing).

Solstices

The maximum distance between the 'Sun' and the 'Earth' that happens twice, once each in south and north of the celestial equator during the annual path of the Earth, around the 'Sun'. These occur on June 21st and December 22nd.

Solubility of a fertilizer nutrient

The quantity of a given nutrient which will be extracted in a specific medium under specified conditions, expressed as a percentage mass of the fertilizer.

Solubility product

For sparingly soluble salts (i.e. those of which the solubility is less than 0.01 mol = dm^{-3}). It is an experimental fact that the product of the total molecular concentrations of the ions is a constant at constant temperature. This product is termed the 'solubility product'.

Solute

The substance being dissolved in a solvent.

Solution

A homogeneous mixture formed by dissolving one or more substance (solid, liquid or gaseous) in another substance (solvent).

Solvent

The substance in which the solute is dissolved.

Sorghum injury

Poisoning of livestock resulting from feeding of young sorghum plants containing Hydrocyanic acid (HCN). HCN concentration is more at seedling stage that is less than 30 days old. This problem is aggravated under drought conditions.

Sorting (seed)

It involves inspection, followed by separating according to variety, size, colour, degree of ripeness, maturity, defects etc. It is actually a separating operation supplementary to grading.

Source strength

The source size times its activity. In calculations, the leaf area index is taken as source size and photosynthesis or net assimilation rate is considered as its activity.

Source

A spot , location or region where there is mass from which the energy is given out into a surrounding system. It may be noted that the objects which are at temperatures of more than absolute zero give out energy.

Southern oscillation

A periodic reversal of the pressure pattern across the Pacific Ocean during EL Nino events.

Sowing

The process of placing seeds in seed bed.

Spacing

The distance between crop rows (inter-row spacing) and between plants within the row (intra-row spacing).

Spatial arrangement

It is defined as the pattern of distribution of plants over the ground, which determines the shape of the area available to the individual plant.

Specialized cropping

It is a cropping plan in which a single crop contributes 50% or more of the total production or monetary receipt (comparable equivalent) in one year.

Species

A unit of classification consisting of a group of closely related individuals which resemble one another in most characters. The members of a species interbreed freely and have a common origin.

Specific crop intensity index

Specific crop intensity index determines the amount of area-time devoted to each crop or group of crops compared to total area-time available to the farmer for crop production during the time period under study.

Specific gravity

The ratio between the weight of a body to the weight of an equal volume of water.

Specific heat

The ratio of the quantity of heat required to raise one gram of a substance by one degree centigrade to that required by one gram of water to raise one degree centigrade. This is also known as specific heat capacity and expressed in joules per kilogram per Kelvin in S.I. units and calories per gram per degree centigrade in C.G.S. system

Specific humidity

The ratio of the mass of water vapour in an air parcel of the atmosphere to the total mass of the moist air. This is expressed as grams of water vapour per kilogram of air.

Specificity

The limitation of a crop to grow in a restricted and definite set of environmental conditions.

Spectral distribution

The distribution of radiant energy in different wavelengths of the solar spectrum.

Spectral solar radiation

The radiation emitted by the Sun, which belongs to selected wavelengths.

Spectrophotometer

An instrument used to measure the intensity of radiation.

Spectrum

In the waveband there are different colours in the visible part. An object can be seen by human eye only when a certain colour is reflected by the same object for which that itself is a source of light, in this connection. One should remember that an image can be seen depending upon the power and frequency of the wavelength. This is not uniform across the total waveband. So, the distribution of power in the solar radiation into different wavelengths is known by this term.

Specular reflection

The reflection of light from a mirror or smooth and non turbulent water surface is non diffusion in nature. These reflections are usually called as specular reflections.

Speed

The time rate of movement of an object.

Spherical pyradiometer

An instrument used to measure the total solar and terrestrial radiation falling from a solid angle.

Spill ways

A conduit in or around a dam for the escape of excess water.

Split plot design

A layout in which one set of treatments is assigned to large plots called main plots into which a replicate is divided and another set of treatments to sub-divisions of the main plots called sub-plots is termed a split plot design. In this sub plot treatment is more the precision of compared to main plot treatment.

Spot treatment

Application of sprays to localized or restricted areas, as differentiated from overall broadcost, or complete coverage.

Spray drift

The movement of airborne spray particles from the spray nozzle beyond the intended contact area. This problem crops up when wind velocity is > 10 Kmph. If a crop is sentive to a herbicide located adjacent to area being sprayed, it will be damaged. Example cotton crop is very sensitive to 2,4-D spray drift.

Sprayer

A machine to apply fluids in the form of droplets of effective size and distribute them uniformly over the plants.

Springging

Vegetative propagation by planting stolons or rhizomes (springs) in furrows or holes in the soil.

Spring

In hydrological terms, an opening on the surface of the earth from which ground water flows. When soils are opened they may occur. In some places natural springs are also noticed.

Sprinkler irrigation

Irrigation by means of sprinkler spaced at intervals on a pipe and is practiced under undulating topography, where water is scarce and cash crops are grown. Three types of sprinkler irrigation are, overhead, perforated type and fixed jets.

Sprout

To put forth new growth from seeds or seed pieces, e.g. potato sprouts, or onion sprouts.

Spudding

Removal of weeds by cutting off below the soil surface.

Spur terrace

A short terrace used to collect or divert runoff.

Squall line or line squall

At different locations of the world this term has been interpreted in different ways. Irrespective of its usage, the central meaning is that the neatly displayed lines of thunder storms or their associated cumulonimbus clouds.

Squall

A strong and sudden onset of wind which decreases later in minutes.

St.Elmo's fire

A luminous, audible and sudden electric discharge that occurs during stormy weather.

Stability

In agrometeorology, this term is often used to refer to the state of the atmosphere. In a given situation, when a parcel of air is displaced, if it offers resistance to move vertically or otherwise and reaches to its original position then the condition is said to be 'stable'.

Stable air

A condition of the air mass in which no disturbance grows and static stability prevails.

Stable

The condition in the atmosphere wherein the decrease in temperature with height is lesser than dry adiabatic lapse rate. Under these circumstances the vertical motions are restricted, consequently precipitation shall not occur.

Stagnation

On certain occasions, the atmosphere exhibits slow or limited dispersion. However, such a situation persists only for time periods of few days. When it happens on larger areas the situation is termed as 'stagnation' in atmosphere.

Stand

A general term for an aggregation of plants with more or less uniformity in composition and habitat conditions.

Standard atmosphere

At the latitude of 45 degrees and at a temperature of 15 degrees centigrade, the normal pressure is given as 1013.2 millibars. This is equal to 29.92 inches or 760 millimeters of mercury at mean sea level.

Standard atmospheric pressure

This can be expressed in the following way. The pressure exerted by 760 millimetres of mercury with gravitational force of 980.70 centimetres per second per second which is corresponding to 1.013×10^6 dynes per centimetre per centimetre.

Standard deviation

- A measure of dispersion of variables around the general mean and is calculated by summing up the squares of the deviation of each observation from the mean and dividing by the number of observations and then extracting the square root.
- Root mean square deviation of the observed values from their average.

Standard score

A score which represents the deviation of a specific score from the mean expressed in standard deviations units.

Standard time of observation

As per the IMD guidelines, the exact time fixed to record the data of the weather elements in any of the observatories.

Standing cloud

Any type of isolated cloud, formed over peaks or ridges of mountains that appears stationary. Altocumulus, Lenticulars are other related terms.

Standing crop

The total amount of biomass of a crop of one or more varieties within an area.

Standing wave

An atmospheric wave that is stationary with respect to the medium in which it is embedded.

States of matter

A substance with mass, in this universe shall be, basically either in solid, liquid or gaseous form. However, some more combinations are artificially possible. In meteorology, the above three basic forms are known as states of matter.

Static model

A model that does not contain time as a variable. In other words, time shall not be used or sidelined in these models. These models are considered as exactly opposite to that of dynamic models. Also see 'Dynamic models'.

Station pressure

It is a fact that the atmospheric pressure changes with the height above the mean sea level. The station pressure is the pressure of the atmosphere before it is corrected to the sea level.

Stefan-Boltzman constant

The universal proportionality constant in the Stefan Boltzman's law, which is equal to 5.77×10^{-5} ergs per centimetre per centimetre per second per degree to the power of minus four.

Stefon-Boltzman law

One of the laws of radiation which states that, the amount of energy given out per unit time from a unit area of an ideal black body is directly proportional to the fouth power of its absolute temperature. Also see 'Absolute temperature'.

Steridian

The unit of solid angle.

Stevenson screen

A wooden box, in which the wooden panes are obliquely arranged and that houses the thermometers, and kept at four feet height above the ground in a meteorological observatory. This is initially used to measure air temperature from the thermometers kept inside, but later on double Stevenson screens have come and these are used to keep thermograph and hygrograph also.

Stomatal transpiration

The most usual method of transpiration, in which water evaporates through stomata. Stomata are small openings which are present on green parts of the plants, mainly on leaves.

Storm

A type of weather system in which strong surface winds blow with a speed between 53 and 63 knots i.e., 64 to 74 miles per hour. Number 11 on Beufort scale wind force.

Straight fertilizer

Quantification generally given to a nitrogen, phosphorus or potassium fertilizer having a declarable content of one primary nutrient only.

Stratocumuls

A low level cloud and the mean height is 0-2 kilometres. 'Stratos' refers to 'layer' and cumulus means 'heap'. These clouds are white or grey in colour in the form of layer, and often seen within a cyclone.

Stratopause

The transition layer between the stratosphere and the mesosphere.

Stratosphere

The layer of atmosphere above the troposphere and tropopause. The lower part of stratosphere is isothermal which means that the temperature remain constant but the temperature increases with height beyond 20 kilometers. This is relatively a quiet layer as compared to troposphere.

Stratus

A low level cloud and the mean height is 0-2 kilometres. 'Stratos' refers to 'layer'. These are grey in colour with almost uniform bottom and top and may give drizzle, ice prisms and snow grains. When the Sun is seen via the cloud its outline is discernible.

Streamline

A more common term used in the aerodynamics of crop plants. A line parallel to the flow of fluid is known by this term. A fluid may be either wind or water.

Stress

A potentially injurious force or pressure acting on the plant which may lead to injury or death.

Strip cropping

Growing soil conserving and soil depleting crops in alternate strips running perpendicular to the slope of the land or to the direction of prevailing winds for the purpose of reducing erosion.

Strip intercropping

Growing two or more crops simultaneously in different strips wide enough to permit independent cultivation but narrow enough for the crop to interact agronomically.

Strip plot design

Two different sets of treatments can be tried in large plot with one set of plots superimposed over the other set at right angles. A block may be divided into strips in one direction to be allotted to one set of treatments and into another set of strips to be alloted to the second set of treatments. This design is used when two sets of treatments need larger plots.

Strip tillage

A tillage system in which only isolated bands of soil are tilled.

Strong breeze

Air in horizontal motion with speed between twenty two to twenty six nautical miles per hour or six on Beufort scale wind force.

Stubble mulching

Stirring the soil with implements that leaves considerable part of the vegetative material or crop residues or vegetative litter on the surface as a protection against erosion and for conserving moisture by favouring infiltration and reducing evaporation.

Stubble

The basal portion of the stems of plants left standing after cutting or harvesting the crop.

Sub soil placement

Placement of fertilizers in the subsoil with the help of heavy power machinery usually followed in humid and sub-humid regions.

Subhumid

This term refers to climatic regions where the moisture conditions range from 30 centimetres in cooler and 90 centimetres in the hotter places. In these areas, according to several climatologists and botanists, the natural vegetation consists mainly of tall grasses and the crops are grown under dry farming situations.

Sublimation

The transition of water from vapour phase to solid phase without passing through liquid phase. The transition of a solid direct into vapour and subsequent condensation without intervening melting stage is also known by this term.

Subplot

A sub division of an experimental plot or sample plot.

Subpolar

The region bordering the polar region, between 50° and 70° north and south latitude on both sides of the equator. This is generally an area of semi-permanent low pressure that exists and where the Aleutian and Icelandic lows may be found. However, a dome of high pressure may form over the cold continental surfaces during the winter, for example, the north American high and the Siberian high.

Subsidence

A sinking or downward motion of air, often seen in anticyclones. It is most prevalent when there is colder, denser air aloft. It is often used to imply the opposite of atmospheric convection.

Subsistence crop

The crop which is grown under certain problematic conditions when no other crop can be grown as floating-rice in flood prone areas.

Subsistence farming

It is farming enterprise which provides food and commodities just sufficient for the farming family, and there are no surplus to sell.

Subsoil

That part of the soil below plough depth or below 'A' horizon.

Subsoiling

Breaking the compact subsoils without inverting them with a special narrow cultivator, showel or chisel at depth–usually from 30-60 cm from the surface.

Substitution cropping

It is substitution of an existing inefficient crop with an identified efficient crop in a zone or region.

Subsurface irrigation

Irrigation of crops by applying water below surface of the ground through pipes.

Subtropical

The region between tropics and temperate zones. This region is characterized by distinct summers and winters. The heat is more in these areas as compared to temperature zone.

Sulphur coated urea (SCU)

It has been developed in recent years as a slow release nitrogen fertilizer and is basically urea with a coating of elemental sulphur, including a binding agent, a sealant and a microbicide. The N content of SCU ranges from 10-37% depending on the thickness of the sulphur coating.

Sulphur deficiency

Symptoms are chlorosis of younger leaves followed by yellowing of older leaves, stunted growth and reduced tillering. Sulphur deficiency affects the whole plant where as nitrogen deficiency affects older leaves. Sulfur deficiency occurs in soil low in organic matter in humid regions. Flooding aggravates sulphur deficiency by converting soluble sulphates in to insoluble sulphides.

Sum of squares

Deviations from the mean, squared and summed.

Summation of temperature

The summing of effective temperatures or growing degree days, for a phenophase or whole of crop growth during its development.

Summer season

A typical mesometeorological term used to denote the monsoon winds with oceanic origin blowing in summer months.

Summer

The period between the summer solstice and the autumnal equinox.

Sun plant

A crop plant that grows well in full sunlight. Otherwise called as heliophytes.

Sun spot

Of the different layers of the Sun, photosphere is the one wherein some non-luminant spots are observed. Under close observation the dim and dark areas which are seen as spots are known as 'sun spots' and one of the main reasons for the existence of these spots is, lower temperatures in these areas as compared to the other areas.

Suncuring

The process of drying in sun e.g. sun-curing of tobacco.

Sunrise

The daily appearance of the sun on the eastern horizon as a result of the earth's rotation.

Sunset

The daily disappearance of the sun below the western horizon as a result of the earth's rotation.

Super adiabatic lapse rate

Almost reverse conditions exist as compared to stable atmosphere. In this situation, the rate of fall of temperature with height in the atmosphere is greater than dry adiabatic lapse rate.

Super cooling

The reduction of the temperature of any liquid below the melting point of that substance's solid phase.

Superphosphate (single ordinary super phosphate)

A commercial product obtained by treating phosphate rock with sulphuric acid and containing about 16 per cent of P_2O_5 mainly water-soluble, along with calcium sulphate and other products of reaction it also supplies calcium and sulphur besides phosphorus.

Supplemental irrigation

The watering of crops in regions where normal rainfall ordinary supplies most of the moisture. It is provided when crop is facing moisture stress due to dry spells.

Supplemental pasture

Additional pastures for use in adverse weather conditions usually annual forage crops for dry periods or winter time.

Supplementary observation

The observations recorded for over and above the routine mandatory weather elements in an observatory. These are required for the study of the condition of the atmosphere under special circumstances like movement of a tropical cyclone, tornado development, etc.

Support crops

Certain fast growing crops which work as support to vine crops e.g. castor or Sesbania sp to support betelvine.

Surface albedo

The ratio of the reflected radiation to the incident electromagnetic radiation at the surface of the Earth. This term is expressed in percentage ratio.

Surface irrigation

A broad group of an irrigation methods wherein water is applied directly on the soil surface from a channel located on the upper side of the field. The water spreads over the fields by gravity of flow.

Surface layer

In micrometeorology, the ecosphere of a crop canopy is conveniently divided into different layers. A surface layer is the one in which different parameters of weather including their energy fluxes are comparatively uniform. The depth of this layer depends upon different factors such as plant height, speed and direction of the wind, angle of the sunrays, etc.

Surface tension

The free energy in a liquid surfaces produced by underlying molecules upon the layers of molecules at the surface. For calculation purposes a hypothetical tension is assumed acting in all directions parallel to the surface and equal to the free surface energy. This mechanical equivalent of free surface energy is called surface tension.

Surface tillage

To till surface without turning the tilled layers.

Surface wind

This is the persistent downslope flow of wind, which dominates the annual and seasonal wind both in speed and direction. The characteristics include its gravitational and thermal origin, relatively smaller area and shallow deep (one hundred to two hundred metres). The surface wind is decoupled from the synoptic gradient flow aloft. The air blowing in horizontal direction near the Earth station and this is measured in open situation at 10 metres height above the ground.

Surfactant

A material which in pesticide formulation imparts emulsifiability, spreading, wetting, dispersability or other surface modifying properties.

Swamp

A completely or partially vegetated area subject to permanent inundation in stagnent or slowly flowing water.

Sylvipastoral system

It is a land management system in which forests are managed for the production of wood as well as for the rearing of domesticated animals. In this system, animals are kept and permitted to graze within the forest. This system therefore, distinguished from other system in which forage (either herbaceous or shrubby) is grown in mixture with forest trees, but is harvested for feeding else where. These latter systems are called agrisilvicultural systems.

Symbiosis

An intimate physiological association of two or more species resulting in mutual benefit e.g. Rhizobium bacteria on the roots of legumes or fern azolla association.

Symbiotic nitrogen fixation

Fixation of atmospheric nitrogen by leguminous plants with the help of root nodule bacteria (Rhizobium spp.).

Synoptic chart

A map or chart that depicts meteorological or atmospheric conditions over a large area at any given time.

Synoptic circulation

The synoptic climatology is a valuable non-quantitative approach to climate description, emphasizing non-periodic weather episodes associated with transient weather disturbances and other circulation patterns at various scales and durations. The synoptic circulations can be tracked on synoptic chart.

Synoptic meteorology

A branch of meteorology concerned with the study of atmospheric phenomena, based on the analysis of charts on which synoptic observations are plotted, for the purpose of weather analysis and forecasting.

Synoptic observations

The observations recorded simultaneously at required number of meteorological stations alongwith meteorological observations. These are useful to know the state of the atmosphere at a given time and also to track the movement of cyclone and monsoon winds. Also, over a relatively small area, some convective storms, etc., can also be observed.

Synoptic scale

The size of migratory high and low pressure systems of the lower troposphere, the cover of a horizontal area of several hundred miles or more.

Related terms: Macroscale, mesoscale, and storms

Synoptic station

A meteorological station in which synoptic data is observed, recorded and sent either to the regional head quarters or to central head quarters, as per the direction of the IMD. The data thus recorded are useful in plotting the isobars, isotherms, troughs, etc.

System

A limited part of reality that contains interrelated elements.

Systemic or translocated herbicide

Herbicide capable of moving within the plant to exert effects throughout the entire plant system irrespective of its place of entry e.g. 2, 4-D.

Systems approach

Studying a system as an entity made up of all its components and their interrelationships, together with relationships between the systems and its environment

T

T.H.I.

An abbreviated form for Temperature Humidity Index.

Tabi

Synonym of *Rabi* season. See *'Rabi'*.

Tagging (labelling)

Adding of a radioactive isotope to the reacting elements or compounds in a biochemical process.

Tail wind

The wind that assists the further movement of an object by blowing in the same direction in which the object is actually moving.

Temperate climate

The climate which includes the warm damp summers and mild wet winters, that are experienced in mid latitudes.

Temperate zone

The portions of the Earth in both the hemispheres between the tropics and polar circles i.e., 23° - 27° from the poles.

Temperature inversion

A situation in which there is an increase of air temperature with increase in height from surface. So, this is the vertical temperature distribution which is contrary to the normal lapse rate.

Temperature lapse rate

The decrease in temperature with increase in altitude in the air is called as vertical temperature gradient. The actual figure obtained from observations of the vertical temperature gradient are known as lapse rate which is estimated to be 6.5 degrees centigrade per kilometer. In other words, the temperature lapse rate is the rate of fall of temperature with increase in height in the temperature.

Temperature of the soil surface

The temperature recorded on a thermometer by placing the bulb just in contact with the surface of the soil after protecting it by shelter from the direct solar radiation.

Temperature profile

In micrometeorology, an expression to represent the temperature of either air, leaf, soil, etc., as a function of height above a surface. Also see 'Profile'.

Temperature range

On any day or for 24 hour period the difference between maximum and minimum temperatures recorded as per the IMD guidelines. For some specific calculations and uses, the temperature range is taken as the difference between the highest and lowest mean temperatures during a given period.

Temperature

- The measure of speed per molecule of all the molecules of a body.
- The concentration of sensible heat in a body measured according to an arbitrary scale such as the centigrade.
- The property of an object measured by the ability to transfer heat to another object. In almost all situations in micro-meteorology any one or all the above can be used for satisfactory explanation of any event in the crops.

Temporary wilting
Wilting condition of the plant due to loss of water which would recover from wilted condition during the night without addition of any water.

Tensiometer
A device for measuring the negative pressure (or tension) of water in soil in situ consisting a porous permeable ceramic cup connected through a tube to a manometer or vacuum gauge. It is used in scheduling of irrigation.

Tephigram
A thermodynamic diagram consisting of temperature and potential temperature on different axes.

Terrace interval
Distance measured either vertically or horizontally between corresponding points on two adjacent terraces.

Terrestrial ecology
According to agrometeorologists this is the study of cropland, grassland, forest, desert, etc., in relation to ecosystems.

Terrestrial radiation balance
See 'Net terrestrial radiation'.

Terrestrial radiation
The infrared radiation emitted by the Earth and its atmosphere including material at terrestrial surface. The wavelength interval ranges from 4 to 120 microns, with its peak at 10 microns on the surface of the Earth. In this context the infrared radiation means the long wave radiation only.

Terrestrial weeds
Weeds which normally start and complete their life cycle on land.

Test of significance
Statistical test designed to distiguish differences due to sampling error from differences due to discrepency between observation and hypothesis.

Thaw

The liquification of snow, ice or any solid state of water at the surface of the Earth because of rise in temperature above zero degree centigrade.

Theodolite

An optical tracking telescope used to observe the motion of a pilot balloon.

Thermal belt

A well defined and earmarked area in mountainous regions in which the vegetation is exceptionally free from abnormal weather conditions like frost, etc.

Thermal capacity

The quantity of heat in joules necessary to raise the temperature of a body or system through one degree Kelvin.

Thermal conduction in a gas

In a gas the phenomenon of thermal conductivity is explained like viscosity. In a column of air the energy varies from layer to layer, which clearly indicates that the thermal conductivity is due to the 'transport of energy'. While conducting heat, the heat causes an equality of temperature through the whole mass of gas under consideration. This is the phenomenon of thermal conduction in a gas.

Thermal conduction in solids

Thermal conduction in a solid is measured by stating the thermal conductivity, which in turn is the ratio of the steady state heat flow along a long rod to the temperature gradient along the long rod. This varies among different types of solids and depends on temperature and purity.

Thermal constants

The sum of growing degree days of temperature that is required for a crop plant to complete its duration after sowing.

Thermal convection

Of the three main processes of energy transfer, the convection is the process in which the heat energy is transferred through a liquid or a gas by the actual movement of molecules of the medium. The same term is used by micrometeorologists to refer to the buoyant convection driven by temperature differences, in crop and also in energy balance equations.

Thermal diffusivity

The thermal conductivity divided by the product of the density and specific heat capacity of a substance is called as thermal diffusivity. Here, the thermal conductivity is a quantity defining the ease with which a substance propagates temperature difference. So, thermal diffusivity can be stated as a measure of facility with which a substance will undergo temperature change.

Thermal equilibrium

If all parts of a system attain the same temperature or are separated by a thermally non-conducting boundary, so that no heat is transferred, then the system is said to be in thermal equilibrium.

Thermal induction

The change in growth and development of plants brought about by a given temperature exposure, e.g. vernalisation.

Thermal low

An area of low pressure due to the high temperatures caused by intensive heating at the surface which it tends to remain stationary over its source area.

Thermal radiation

Same as terrestrial radiation. Also see 'Long wave radiation'.

Thermal stratification

The conditions of a body of water or air in a crop in which the successive layers have different temperatures, each layer more or less sharply differentiated from the adjacent ones, the warmest being at the top.

Thermal velocity

In a given fluid the weight of a body is balanced by its drag. Under this situation, the speed at which the weight of the body is balanced is known as thermal velocity. This is also called as 'fall speed'.

Thermal wind

The difference in temperature in the geostrophic wind between two heights in the atmosphere over a given position on the Earth. In other words, the thermal wind is the vertical shear of geostrophic wind.

Thermal

In meteorology, an area of rising air mass. Also, any property explained with temperature or heat energy as main quantity, in any phenomena.

Thermistor

- In a semi conductor the electrical resistance of which may be of the order in such a way that it can exhibit rapid and extremely large changes in resistance for relatively small changes in temperature. These changes vary in a known manner with temperature.
- A thermally sensitive resistor whose primary function is to exhibit a change in electrical resistance with a change in body temperature.

Thermocline

The layer in a thermally stratified body of air or water within which the temperature decreases rapidly with increasing depth, usually at a rate greater than 1 °C per metre of depth or distance in vertical direction.

Thermocouple

A pair of dissimilar metals producing a potential difference when their junction is heated. In this device the conversion of thermal energy into electrical energy takes place at a known rate. Thermocouples made up of different metals like iron, copper, zinc, etc., are used in crop canopies to record the temperature differences and the use of the metal is specific for each situation.

Thermodynamic diagram

A diagram in which various thermodynamic processes are depicted graphically as a path. The axes are thermodynamic parameters of particular working substance. So, the area represents energy, eg. Pressure-volume-temperature diagram, temperature-entropy diagram, etc.

Thermodynamic frequency

The study of the general laws governing processes which involve heat changes and the conservation of energy is called thermodynamics. The thermodynamic frequency which is also known as thermodynamic probability is the number of microstates corresponding to any given macrostate, in a thermodynamic system.

Thermodynamic laws

There are three laws of thermodynamics, of which one is explained in this book under 'First law of thermodynamics'. The second law states that, the spontaneous transformation of energy is accompanied by dispersal of a part into non available heat. For example, in respiration in plants. The second law can also be stated as that the heat cannot flow from a cold to a hot body without the performance of work by some external agency. The third law states that, the absolute zero temperature is not attainable.

Thermodynamic reference process

To study the complex atmospheric realities, different physical processes and principles are being used by the scientists in different localities of the world. In a thermodynamic process like an adiabatic process or a series of thermodynamic states of an air parcel are used as reference to arrive at a conclusion for a complex process then the process that is taken for comparison is called as thermodynamic reference process.

Thermodynamic scale

A temperature scale developed by Lord Kelvin which is independent of the properties of any matter. This is also known as Kelvin scale of temperature or absolute scale of temperature.

Thermodynamic wet bulb temperature

The maximum possible lowest temperature to which an air parcel can be brought to coolness by adiabatic evaporation of water.

Thermodynamics

Science that discusses the relation between the heat energy and mechanical work. This science, with its laws, perfected the equivalence between the work done and the heat produced. By purely logical reasoning the laws of other sciences can be deduced. This science is based upon common observations and the principles give the relation of heat to other forms of energy like electrical, chemical, radiation, etc. The laws basically govern the conversion of energy from one form to another.

Thermograph

A continuously recording thermometer which exhibits the time variation of temperature on a graph.

Thermometer

An instrument used to measure the temperature. If a thermometer is used to measure the temperature of the soil it is called as soil thermometer. Likewise, depending upon the medium of measurement the names are given.

Thermometric conductivity

The ratio of the coefficient of thermal conductivity to the thermal capacity per unit volume of the material. This is also called as thermal diffusivity.

Thermoperiodicity

Characteristics of a plant to exhibit its behaviour with reference to variations in air temperature because of differences in exposure to light and dark periods.

Thermopile

An instrument used to detect and measure thermal energy, i.e., heat radiation. It consists of one or more pairs of thermocouples connected either in a series or in parallel, from which the thermal energy is converted directly into electrical energy.

Thermosphere

The atmospheric layer above the mesosphere in which there is a rapid increase in temperature with height. There is a constant low temperature at about 80 kilometres. From there onwards, the values go to 200 degrees Fahrenheit and further upto an unidentified upper limit.

Thermostat

A device used for automatic regulation of temperature which usually consists of a bimetallic strip.

Thickness

The vertical depth of an air mass, across which the pressure is uniform.

Thinning

An agronomic practice of removing extra plants from thickly populated crop stand with an idea of maintaining optimum plant population per unit area.

Thornthwaite system

A system of climatic classification proposed by Thornthwaite.

Three dimensional diagrams

A form of graphical representation of numerical data through diagrams which are in the form of cubes. The volume of the cubes are proportional to the magnitude of the variate.

Threshing (soil)

The movement of soil in any direction as a result of kinetic energy imparted to the soil by tillage tool.

Threshing (crop) :

The process of separating grains/pods from a crop by manually or mechanically.

Threshold temperature

The temperature below which no plant physiological process takes place in a crop plant.

Thunder sky

Sky covered with cumulonimbus clouds producing thunders.

Thunder storm

A cumulonimbus induced storm accompanied by violent wind, rain, lightning, thunder, etc. The vertical area occupied by this storm depend on the intensity of the ascending air currents. The facts influencing its development are atmospheric instability, supply of warm moist air, lifting of potentially unstable air, etc.

Thunder

The audible noise resulting from the explosive effect of electric discharge caused by very rapid expansion and contraction of air within a cloud. This is usually associated with a cumulonimbus clouds during the lightning discharge. Because light waves travel faster than sound waves, the lightning is seen before a thunder is heard.

Tide

The periodic rising and falling of the earth's oceans.

Tile drainage

Pipe made of burned clay, concrete or ceramic material in short lengths, usually laid with open joints at suitable spacing and depths in the soil or subsoil to collect and carry excess water from the soil.

Till

To plough or cultivate soil for seeding.

Tillability

The degree of ease with which a soil may be manipulated for a specific purpose.

Tillage action

The action of a tillage in executing a specific form of soil manipulation such as soil cutting, pulverizing or inversion.

Tillage depth

Vertical distance from the initial soil surface to a specific point of penetration of the tool.

Tillage

The use of implements for mechanical manipulation or ploughing the land to prepare seedbeds conducive for field crop production.

Tiller

Side-shoot growing from the base of the stem (at ground level) of a cereal or grass plant.

Tillering stage

The appearance of the first tiller from the axillary bud in one of the lower most nodes. It is characteristic feature of crops belongs to the family gramineae with few exceptions.

Tilt

The inclination to the vertical of a significant feature of the pressure pattern.

Tilth

The physical condition of the soil with respect to its fitness for the planting germination or growth of a crop.

Top dressing

The application of manure or commercial fertilizer to a crop that is already established.

Topoclimatology

A study of the climate of a locality or of a small area as influenced by its slope, surface, vegetation and other characteristics with reference to its topography. The main emphasis of the study includes the configuration of the area and surface features.

Topography

A technique to describe the surface features of the soil, land or the Earth for different purposes in explaining various meteorological phenomena.

Topping (tobacco)

The removal of the terminal bud with or without some of the small top leaves just before or after the emergence of a flower bud.

Tornado

This is a violent, destructive storm of small horizontal dimensions. A cumulonimbus cloud forms into a funnel shape with a vortex extending from the base of the storm to the surface. The whirl wind encircles a small dimension of about 500 metres. These are capable of causing severe structural and other damages. The violent winds associated with this abnormality are strong upward air currents.

Torrential rain

A rapid and violent heavy rainfall effecting economic losses, occasionally.

Total internal reflection

The reflection of a light ray from a surface at such an angle that the refracted ray must be bent away at an angle greater than the critical angle. Under this situation, the light cannot emerge at all and is internal reflected totally.

Total lift

The upward force induced by the gas or gaseous mixture in a balloon which is equal to the free lift plus the weight of the balloon and the attached equipment.

Total nomadism

In this systems, the animal owners do not have a permanent place of residence. They do not practice regular cultivation, and their families move with their herds.

Total potential energy

In meteorology, the internal and gravitational potential energies of a column of air in the atmosphere. In micrometeorology of crop plants, the gravitational and potential energies of relatively small parcels of eddies.

Total radiation pyrometer

An instrument used to measure the total radiation.

Total radiation

The sum of solar radiation and terrestrial radiation.

Totalizing anemometer
An instrument used to integrate the air movement directly.

Trace gases
The atmospheric constituent gases other than oxygen, nitrogen and water vapour. This version is holding good as far as upper air circulations are concerned. For precise usage in other branches like agroclimatology and micrometeorology this term is used for a minor constituent category in a comparatively major part of gases.

Trace
An unmeasurable or insignificant quantity of precipitation amount (less than 0.005 inch).

Trade winds
This word is derived from 'tread', which was later changed to the present usage 'trade'. These are conspicuously reliable and steady winds which are helpful in shipping commerce, etc. When the equatorial region gets heated, the air rise from the surface and passes to the upper layers. The pressure of the atmosphere near the surface decreases in due course. Air move towards this low pressure area from both north and south and this phenomenon continues right through the year. The resulting winds also take the same course or track and are called trade winds. These are also called as tropical easterlies. Because of the rotation of the Earth, when these winds are deflected north of the equator, they are called northeast tradewinds and when deflected to south of the equator, they are called southeast trade winds. The main feature is that these winds blow in the tropics that too in the low troposphere.

Trail (experiment)
A group of plots to which treatments have been imposed, arranged in a manner dictated by the chosen experimental design.

Trajectory
In meteorology, this can be referred to the curve or path in space, traced by a moving air parcel. It can be assumed that, the tracing of points successively occupied by a parcel of air are said to be a path, in this context. However, in different branches of science there are different meanings for this term, e.g. the path of a projectile is also called by this term.

Translucent body

An object through which a part of light can pass but through which objects cannot be seen clearly, eg., frosted glass.

Transmission coefficient

The fraction of incident radiation intensity transmitted by a source.

Transmissivity

In micrometeorlogy, this can be stated as the ratio of transmitted to the incident electromagnetic radiation on a crop canopy. In mesoscale and macroscale meteorology, this is a measure of luminous flux remaining in a radiation beam after it passed through a particular distance in the atmosphere.

Transmissometer

An instrument used to measure 'extinction' of light segment of the solar radiation in a particular atmospheric path.

Transparent bodies

Those that permit the passage of light from objects in such a way that the objects can be seen distinctly through the substance.

Transpiration coefficient

The loss of water in vapour form from a crop plant through the stomata and lenticels is called transpiration. The ratio of the oven dry material produced by a plant during the growth period to the total amount of water transpired during the same period is termed as transpiration coefficient.

Transpiration efficiency

The quantity of dry matter produced by a crop plant for every kilogram of water transpired. This is measured in grams.

Transpiration pull

A capillary pull arising out of a fall in pressure as a result of transpiration at the top of the plant. It is an unavoidable loss of water demanded by atmosphere.

Transpiration ratio

The quantity of water transpired to produce a unit amount of dry matter. Generally weeds have more transpiration ratio than crop plants.

Transpiration

The process in which the water in plant tissues is released as water vapour into the air surrounding the same.

Transplanter

A machine which sets seedings into the ground in rows. The spacing between the seedlings is adjustable.

Transplanting

The process of planting seedlings in prepared seed beds in main field.

Transport phenomenon

The kinetic theory of matter states that, the particles of matter in all states of aggregation are in vigorous motion. Likewise, for gases the phenomenon that can be explained on the basis of movement of gas molecules is known as transport phenomenon. As per the kinetic theory of gases a gas has a tendency to reach a steady state or equilibrium state because the theory states that the molecule of a gas are in a constant state of thermal agitation. When a gas is not in equilibrium state the major phenomena, i.e., viscosity, thermal conductivity and diffusion occur. So, in micrometeorology almost all important physical properties refer to non-equilibrium state only.

Transported soil

Soil formed by the consequent or subsequent weathering of materials transported and deposited by some agency such as water, air etc.

Transverse drainage

A method of drainage where the drains are placed in a direction more or less at right angles to the direction of the steepest slope of the land to be drained.

Transverse waves

The waves in which the vibration or displacement takes place in a plane right angles to the motion of the wave.

Treatment

A single state of some factor being varied in an experiment, such as a rate of herbicide, method of applying a herbicide, sources of fertilizer etc.

Tree cropping

It is based on the concept of using tree crops as the basis of an agricultural system. It entails replacement of grain crops with tree crops, consisting of an upper tree story of one or more type of forage trees, a ground level permanent pasture and one or more types of domestic animals that feed on the pasture or on tree products.

Trickle irrigation

A method of applying water directly near the root zone of the plants through a number of low flow-rate outlets generally placed at short intervals along small tubing. Some times it is referred to as 'drip irrigation', followed for high value crops in arid regions. It records highest water use efficiency compared other methods of irrigation developed first in Isreal.

Triple cropping

Growing three crops a year in sequence.

Triple factor

A factor, the change of which sets off a chain reaction in an ecosystem.

Triple superphosphate

A commercial product obtained by treating phosphate rock with phosphoric acid and containing about 46 per cent P_2O_5, mainly water-soluble.

Tripplepoint temperature

The temperature at which all the three phases of a substance exists in equilibrium , at exactly under one pressure. It is 273.16 degrees Kelvin for water and is the basis for Kelvin scale.

Tropic of cancer

The most northern point on the earth where the sun is directly overhead, (23° 51' degrees north latitude).

Tropic of Capricorn

The most southern point on the earth where the sun is directly overhead, (23°.5[1] degrees south latitude).

Tropical air masses

These are low latitude air masses. There are three types of tropical air masses. 1. Wintertime tropical maritime, the source of which is over the warm oceans, 2. Summertime tropical maritime, the source of which is the belt of the great semipermanent highs of the tropical oceans, and 3. Continental tropical, the source of which is the subtropical high pressure land masses. Among these tropical air masses the common features associated are heavy precipitation, convective instability, violent thunderstorms, etc.

Tropical cyclone

A warm core low pressure system which develops over tropical and subtropical, waters, and has an organized circulation.

Tropical depression

An intense tropical cyclone with maximum winds upto seventeen metres per second. A tropical cyclone is tropical in origin and a size of few hundreds of kilometers. The central region of the cyclone is called the 'eye' of the storm with diameter ranging from 50 to 100 kilometres. The cyclone is accompanied by heavy rains, violent winds and even thunderstorms. The hurricanes and typhoons are the most severe cyclonic storms.

Tropical disturbance

Rotary circulation, slight or absent at surface but sometimes better developed aloft. No closed isobars and no strong winds are observed. This is a common phenomenon in the tropics.

Tropical mass

A large mass of air originating near the tropical torrid zone on the Earth.

Tropical storm

Same as tropical depression or tropical cyclone in which the maximum speed of the wind is between 17 and 32 miles per second.

Tropical wave

An easterly wave, with relatively low pressure moving westward through the trade wind easterlies.

Tropics

There are tropics of cancer and Capricorn respectively at 23 degrees 30' N and 23 degrees 30' S of the equator on the globe. These are the imaginary limits over which the Sun is overhead directly atleast once in a year. There is a lot of significance for these areas on the Earth's surface with reference to formation of equatorial calms, trade winds, etc. In the Indian sub-continent the behaviour of monsoon is predicted in most of the theories based on the conditions existing over these areas.

Tropopause

The boundary layer between the troposphere and stratosphare. This is a shallow layer of about 15 kilometres on an average above the sea level and vary in elevation from 8 kilometres to 1 kilometres near the poles and the equator respectively. This layer is characterized by sharp temperature inversion, vertical turbulence and abrupt change in lapse rate.

Troposphere

The lower layer of the atmosphere between the surface of the Earth and stratosphere. In this layer clouds form, because of the fact that 99 percent of the precipitable water is present. About 5 percent of gases of the total atmosphere are present and the average height of this layer is 14 kilometres and at poles it is 8 and at the equator it is 16 kilometres. There is a decrease in temperature with increasing height at a mean lapse rate of 6.5 degrees centigrade per kilometre.

Trough

Region of the atmosphere in which the pressure is low relative to the surrounding regions at the same level. It is represented on a synoptic chart by a system of nearly parallel isobars or contours of approximately v-shape which are concave towards a depression.

Truck crops

Crops which yield in tonnes and grown for distant market requiring heavy transport.

True allelopathy
The active substances may be released as such from same plant

True density
The mass per unit volume of the particles of a material.

True digestibility
Actual digestibility or availability of a feed, forage or nutrient as represented by the balance between intake and faecal loss of the same ingested material. Thus the true digestibility of a feed by animals is the apparent digestibility of the feed plus the unavoidable metabolic excretions.

True solar time
Same as 'Local apparent time'.

True wind
For a moving object true wind is the vector sum of the apparent wind and the velocity of the object.

Truncated soil
Soil which has lost material from the upper part of the profile by the processes of erosion.

T-sonde
A radiosonde used to measure temperature only.

Tsunami (Seismic wave)
An ocean wave with a long period that is formed by an underwater earthquake or landslide, or volcanic eruption. It may travel unnoticed across the ocean for thousands of miles from its point of origin and builds up to great heights over shallower water.

Tuber crop
Crops possessing enlarged underground reproductive stems, roots e.g. potato, cocoyam, sweet potato etc.

Tundra climate / Steppe climate
A type of climate which produces tundra vegetation. It is too cold to support growth of trees but doesn't have a permanent cover of snow or ice.

Turbidity

In hydrological terms, the cloudiness caused by sediment suspended in water. Rivers are generally at their most turbid following a rainstorm.

In agrometeorological terms any condition of the atmosphere which reduces its transparency to radiation, especially to visible part.

The optical depth due to the presence of solid or liquid particles in a vertical column of atmosphere.

Turbulence

An irregular motion of air caused by eddies in contrast to a smooth laminar flow. In fluid flows the apparent chaotic motion caused by instantaneous velocity of fluid elements.

Turbulent flow or eddying flow

See 'Turbulence'.

Turf

The upper stratum of soil filled with the roots and stems of low growing plants especially grasses.

Turgid

Describing a cell or tissue which is firm and plump because of the internal pressure resulting from the osmotic uptake of water.

Turgor pressure

The pressure exerted on the cell wall by the cell contents inside it when the cell contents are fully turgid.

Turn around time

A period between the harvesting of the preceding and planting of the succeeding crop in a specific field.

Twilight (dusk)

The evening period of waning light from the time of sunset to dark.

Twister

Synonym for tornado.

Two stream approximation

The specification of upward and downward radiant 'fluxes' by a single parameter for each direction assuming isotropic radiances in each hemisphere in a plane parallel atmosphere.

Type (plant)

A group of varieties of similar characters or an admixture of seed of indistinguishable varieties but having some characteristics in common, like grain type jowar, yellow maize etc.

Type 1 error

The rejection of hypothesis when it is actually true.

Type II error

The probability of accepting a hypothesis when it is actually false.

Typhoon

An hurricane in the South Chine sea is termed as 'typhoon'. This is a tropical cyclone or storm with maximum winds in excess of 32 metres per second.

U

Ultisols

Soils that are low in bases and have subsurface horizons of alluvial clay accumulations. They are usually moist, but during the warm season of the year some are dry.

Ultraviolet radiation

The part of solar spectrum, having wavelengths shorter than visible segment and range from 0.005 to 0.4 microns. This radiation is detrimental to crop plants and is chemically very active.

Ultraviolet

Shorter form of ultra violet radiation. See 'Ultraviolet radiation'.

Umbra

A region of complete shadow in which there is no light.

Undercast

An opaque cloud layer viewed from an observation point above the layer. From the ground, it is called as an overcast.

Unit value of fertilizer ingredients

One per cent of N, P_2O_5 and K_2O present in one of tonne of a fertilizer is called one unit. The unit value of a plant nutrient in a fetilizer is the price of one tonne of the fertilizer divided by the percentage content of the particular nutrient.

Units of measurement

An abbreviated form for this term is 'unit' which means a quantity or dimension characterized as a specific measuring standard.

Universal gas constant

The constant linking pressure, volume, and absolute temperature of a mole of ideal gas, equivalent to 8.3143 joules per Kelvin per mole or 1.9858 calories per degree Celsius per mole.

Universal Time Coordinate (Zulu, Greenwitch Mean Time)

One of several names for the twenty four hour time.

Unsaponifiable matter

It includes all those constituents of fats which are not saponified by alcoholic caustic potash and which are soluble in pertroleum ether or ethyl ether.

Unsettled weather

Sky with intermediate stages of different forms and types of clouds.

Unstable air

Synonym of 'instability'. A condition of the atmosphere in which static instability exists.

Unstable atmosphere

The condition of the atmosphere in which the decrease in temperature with increase in height from the ground is greater than dry adiabatic lapse rate. In meteorology, the condition of a system which amplifies a particular type of imposed disturbance and the type of unstableness depends on the type of disturbance, e.g. dynamic convection.

Unstable

The condition of a system which implies a particular type of imposed disturbance. The type of instability depends on the type of disturbance, e.g. convection of dynamic and baroclinic natures.

Upland rice

Rice grown in rainfed, naturally well-drained soils, without surface water accumulation, normally without phreatic water supply that are normally bunded or not bunded.

Upper and lower plastic limits

The upper limit is that moisture content at which the soil water will barely flow under an applied force. The lower plastic limit is that moisture content at which the soil can be rolled out into wire.

Upslope effect

The cooling of an air flow as it ascends a hill or mountain slope.

Upslope fog

The fog which is formed when warm, moist surface air is forced up a slope by the wind.

Upwelling

The process by which water rises from a lower to a higher depth.

Upwind (Windward)

The direction from which the wind is blowing.

Urea

A sythetic, non-protein organic compound, crystalline or made into granules or prills for fertilizer use and containing 46 per cent nitrogen.

Urea formaldehyde

Slow release nitrogen fertilizer produced by reaction between urea and formaldehyde.

V

Vacuum

A space in which there is no air or gas. This is more an imaginary situation. This term indicate the air or any other gases at the minimum most possible pressure in a space.

Valley breeze

A warm air mass moving up along the slope of the mountain. These breezes are intense and persistent and often result in orographic rain. Also called as anabatic wind.

Valley wind

The upward flow of wind in the afternoon from a heated valley relative to the existing mountain or hill region.

Vapor trail (condensation trail)

A cloud like streamer or trail often seen behind aircraft flying in clear, cold, humid air. A vapor trail is created when the water vapor from the engine exhaust gases are added to the atmosphere.

Vaporisation

The changing process of a substance from a state of liquid to a state of gas. Also see 'Evaporation'.

Vapour drift

The movement of vapours from the area of application to other areas.

Vapour pressure deficit

The difference between the actual vapour pressure in the crop canopy and the vapour pressure at saturation. This is often considered as an important indicator of dryness in the crop stands. Also see 'Saturation deficit'.

Vapour pressure

- The force per unit area is 'pressure'. Vapour pressure is the partial pressure exerted by the vapour of any evaporating liquid. In meteorology, often, this term is applied to water vapour pressure. In the process of reaching to the higher levels in the atmosphere the water vapour mixes with other gases in the air and exerts pressure on these gases.
- The pressure exerted against the walls of the container and the surface of the water by the water vapour molecules.

Vapour

The gaseous form of a substance at ordinary temperature and used as an abbreviation for water vapour. Also see 'Vapour pressure'.

Variable ceiling

The phenomena which occurs when the height of a ceiling layer increases and decreases rapidly.

Variable

A characteristic or phenomenon which may take on different values.

Variance

Mean squared deviations of variates from their mean. The square of the standard deviation.

Variate

single observation or measurement.

Variation

A deviation from the typical in the anatomical or physiological characteristic of an organism.

Variety

A sub-division or a group of plants within a species which is characterized by growth, plant fruit, seed or other character by which it can be differentiated from other seeds of the same kind.

Vector

Any physical quantity which requires a direction to be stated in order to define it neatly. A vector quantity has both magnitude and direction.

Veering

The clockwise change in the direction of the wind, which implies that, as the direction of the wind is changing the azimuth is increasing.

Vegetation cover

The sum total of plants and plant material such as leaves, stems and reproductive parts that form the coverage on the surface of the soil. It is also used to designate the sum of the living crop plants over an area.

Vegetative cover

A soil cover of plants or grasses and trees irrespective of species.

Vegetation

Sum total of plants in a given area.

Vegetative propagation

The production of a complete plant by asexual methods of grafting, layering, cutting, division, separation etc.,

Vegetative stage

Growth stage of a crop plant prior to flowering or maturity.

Velocity of light

Velocity is the rate of motion in a specific direction and velocity of light is the speed with which light travels, i.e., three lakh kilometers per second (approximately) whch is an universal constant.

Verga

The falling precipitation from a cloud that evaporate before it reaches the ground surface. This is commonly associated with nimbo stratus clouds.

Vernal equinox

Taking place in the northern hemispheric spring, it is the point at which the elliptic intersects the celestial equator. Days and nights are most nearly equal in duration. It falls on or about march 20 and is considered the beginning of spring in the northern hemisphere and autumn in the southern hemisphere. It is the astronomical opposite of the autumnal equinox.

Vernalization

The process to induce flowering in plants by exposing them to low temperature.

Vertical mulching

Incorporation of vegetative thrash in a band in the soil for the purpose of harvesting rain water and conserving soil and water. Generally it is followed for black cotton soils in rainfed areas.

Vertical revolution in agriculture

Maximising production per unit land area per unit of time, with intensive cropping systems, high production inputs and improved management practices.

Vertical temperature

An abbreviated form of vertical temperature variation which indicates that there is a change of temperature with increase in altitude. In the troposphere there is a decrease in temperature with increase in height.

Vertical visibility

The distance an observer can see vertically into an undefined ceiling.

Vertical wind profile

A series of wind direction and wind speed measurements taken at various levels in the atmosphere cropped field that show the wind structure of the atmosphere crop over a specific location.

Vertisols

Clayey soils with high shrink-swell potential that have wide, deep cracks when dry. The clay is of montmorillonite type (2:1 clay mineral). Most of these soils have distinct wet and dry periods throught out the years.

Viability

The capability of a plant structure (seed, cutting etc.) to show living properties like germination and growth.

Vigour

A condition of active good health and material robustness in seeds which upon planting, permits germination to proceed rapidly and to compete under a wide range of environmental conditons.

Virgin soil

A soil that has not been significantly disturbed from its natural environment.

Virtual image

An image seen at a point, like the one seen in a plane mirror from which the rays of light appear to come to the observer but do not actually do so. (On any screen image, cannot be obtained like a real image).

Virtual temperature

The temperature at which air would have the same density as a sample of moist air at the same pressure.

Viscosity

The property of a gas or a liquid where by, they tend to resist the relative motion within themselves. This is possible because of the friction arising from molecular exchange and impact. Often, this is also referred to as molecular viscosity to distinguish it from eddy viscosity. Also see 'Eddy viscosity'.

Visibility

An abbreviated form of visual range and this is the maximum possible distance at which any specified can be seen with a proper background.

Visible radiation

The range from 0.4 to 0.7 microns of electromagnetic radiation which can be seen by human eye. This part of the solar spectrum contains forty one percent of total energy of the spectrum and is very important for all life on the Earth.

Visible solar radiation
See 'Visible radiation'.

Volmet broadcast
Broadcast of meteorological information flying airplanes.

Volt
The S.I. unit of electric potential, potential difference, and electromotive force. This the ratio of one joule of work done to one coulomb of charge moved.

Voltmeter
A device used to measure the potential difference between two points.

Volume weight of soil
It indicates the number of times heavier a given volume of dry soil is than water that will occupy the same total soil volume.

Volume weight
A figure denoting the number of times a dry soil is heavier than an equal volume of water. Usually, the pore space of dry soil is also included in calculations.

Volume
The measure of bulk or space occupied by a body. Expressed in cubic metres, cubic feet, etc.

Volunteer plants
Unwanted plants growig from seed that remains on the field from a previous crop.

Vortex
Any circular or rotary flow in the atmosphere that possesses vorticity.

Vorticity maximum
A center or the maximum of the vorticity field of a fluid.

Vorticity
A measure of rotation of an air or fluid flow. In synoptic meteorlogy, the relative vorticity about the local vertical. In engineering, the swirl and mathematically the curl of the velocity vector.

W

Wage
> The amount paid periodically to the labourer for the work done in the field.

Wall cloud
> An abrupt lowering of a cloud from its parent cloud base, a cumulonimbus or supercell, with no visible precipitation underneath. Forming in the area of a thunderstorm updraft, or inflow area, it exhibits rapid upward movement and cyclonic rotation.

Walter circulation
> A thermally driven longitudinally cellular circulation over equatorial oceans, particularly pacific ocean, Peruvian coast, etc. These circulations rise in the West and sink in the East.

Warm advection
> The horizontal movement of warmer air into a location.

Warm air
> If the temperature of air is relatively warm as compared to an underlying surface or another adjacent air column or within itself, then the air is said to be warm air.

Warm front

A 'front' advancing so that warm air mass replaces cold air mass.

Warm high

A high-pressure system that has its warmest temperatures at the centre of circulation.

Warm low

A low pressure system which has its warmest temperatures at the center of circulation.

Warm season crop

A crop species that makes its major growth during the warmer part of the year.

Warm sector

The state in lower troposphere in between a warm front and the following cold front.

Washout

During summer months, because of expansion of atmospheric gases the minute pollution particles present in the atmosphere reach middle and upper part of the troposphere . These particles again fall on the Earth alongwith the rain drops because of collision and coalescence, which is known as washout. Also see 'Rainout'.

Wastelime (By-product lime)

An industrial waste or by product containing calcium and magnesium in forms that will neutralize acids. It may be designated by prefixing the name of the industry or process by which it is produced; for example gas lime, tanners' lime, lime kiln ash.

Water application efficiency (E_a)

The percentage of water applied that can be accounted for as moisture increase in the crop root zone of the soil. It is given by the formula $E_a = \dfrac{W_s}{W_f} \times 100$.

Where,

W_s = Water stored in the root zone of the plants,

W_f = Water delivered to the field at supply channel.

Water application rate

The rate in cms depth or hectare cm per hour that irrigation water is applied to fields.

Water balance in the soil

The sum of rainfall and irrigation water is equal to the sum changes in soil moisture, evapo-transpiration, percolation, and run-off.

Water balance

It is hydrological balance between raifall plus irrigation to that of changes in soil moisture, evaporation, percolation and runoff.

Water compaction

Water furnished by destruction of pore space owing to compaction of sediments.

Water conservation

Physical control, protection, management, and use of water resources in such a way as to maintain growing crop and forest lands, vegetation cover, wildlife habitat and maximum sustained benefits to people, agriculture, industry, commerce, and other segments of the national economy.

Water conveyance efficiency (E_c)

It is the proportion of water delivered to the irrigated field from the total water diverted from the source. It is given by the formula

$$E_c = \frac{W_f}{W_d} \times 100.$$

Where

W_f = Water delivered to the irrigated plot at the field supply channel.

W_d = Water delivered from the source

Water cycle (Hydrological cycle)

The vertical and horizontal transport of water in all its states between the earth, the atmosphere, and water bodies (seas, oceans, lakes etc).

Water deficit

Shortage of water or inadequate water supply. The amount of water needed to return the soil moisture status to the field capacity (FC).

Water distribution efficiecy (E_d)

The extent to which water is uniformaly distributed in a irrigation system (E_d). It is given by the formula $E_d = \left(1 - \dfrac{\bar{y}}{d}\right) \times 100$.

Where

d = average depth of water stored along the run during the irrigation;

$\bar{y}$ = average numerical deviation from d.

Water equivalent

The depth of water measured from the melting snow or ice assuming the corresponding measurement on a horizontal surface with no infiltration and evaporation.

Water front

The portion of shoreline that has been intensively developed for commercial purpose. The activities include fishing, transportation, etc.

Water furrow

A shallow trench or depression between two raised soil beds in which irrigation water or surface drainage can flow.

Water gates

A water control device, made of wooden planks or metal sheet and used for opening and closing water supply. It is palced in the head ditch near farm or in field ditches.

Water harvesting

Conservation of rain water under unirrigated condition, by collecting runoff of in order to supplement soil moisture as life saving of crop or to irrigate at critical stages of crop in an adjacent area.

Water holding capacity

The weight of water held by a given quantity of absolutely dry soil when saturated. It varies according to type of soil, expressed on volume basis as cm of water per metre depth or inches per feet depth.

Water potential
- The capability of soil water to do work compared with free-water. The water potential at the surface of free-water is taken as zero. It is denoted by (psi) and measured in bars.
- The capacity of soil water compared to free water with particular reference to do the work.

Water quality
It refers to the characterstics of a water supply that will influence its suitability for a specific use, like for irrigation of crops; usually the emphasis is placed on the chemical and physical characteristics of the water. The quality is assessed in terms of salts, carbonates, bicarbonates, sulphates and chlorides etc.

Water regimes
Regime means level; different levels of water (standing or flooding) depth applied in irrigation experiments, such as 2.5, 5, 7.5, 10 cm water to rice crop etc.

Water requirement (WR) of crops
It is that quantity of water regardless of its source, required by a crop in a given period of time for its maturity, and it includes losses due to ET plus the unavoidable losses during application of irrigation water and water required for special operations such as land preparation, puddling, leaching etc., and is expressed in depth for given time. $WR = IR + ER + GWC - Wd$

where

IR	=	Irrigation requirements;
ER	=	Effective rainfall;
GWC	=	Ground water contribution;
Wd	=	Drianage from root zone

Water saturation deficit
Parameter for measuring water deficit in plants.

Water seeded rice
Sowing of soaked rice seed in standing water (7-15 cm depth) on well prepared seedbed.

Water spout

A tornado on an open water surface. The water spouts take out fish, frogs, etc., from sea or river water and throw it on the land mass along with water making it a rain of frogs.

Water stable aggregate

A soil aggregate stable to the action of water such as falling drops or agitations as in wet-sieving analysis.

Water storage efficiency (E_S)

It indicates how completely the water required before irrigation has been stored in the rhizosphere during irrigation. It is given by the formula $Es = \dfrac{W_S}{W_R} \times 100$

where, W_S = Water stored in the rhizosphere after irrigation,

W_R = Water needed in rhizophere prior to irrigation.

Water substance

Water in any of its triple physical states viz., the liquid, the ice, and the vapour.

Water table

The depth below which the ground is saturated with water.

Water vapor

Water in gaseous form.

Water

When the hydrogen and oxygen molecules exist together, through hydrogen bond in a liquid form it is called as water. It is also known as 'universal solvent' and 'fluid of life'. It has molecules which not only cling to each other (cohesion) but also to many different surfaces (adhesion). The water in its solid form is less dense than in its liquid form. It has a high specific heat and high heat of vapoursation (evaporates slowly when heated). It is a fluid, because it takes the shape of the vessel in which it is kept.

Waterlogging

A condition of land where the ground water, soil water content reaches a level that is detrimental to plants and other useful soil micro-flora and fauna.

Watershed management

The planned use of watershed lands in accordance with predetermined objectives, such as the control of erosion, stream flow, sedimentation and improvement of vegetative cover and other related resources.

Watershed

- It is the area above a given point on a stream that contributes water to the flow at that point. Catchment basin or drainage basin are synonymous with it.
- The whole surface drainage area that contributes water to a lake, tank or pond. Recently, watershed technology is gaining priority on the research agenda of all agricultural institutions.

Watershedbased farming system

It involves the optimum utilization of the catchment precipitation through improved water, soil and crop management directly through infiltration of monsoon rainfall after runoff collection and storage the improvement of agriculture on the watershed.

Water yield

It is the total annual runoff volume.

Watt

The S.I unit of power equivalent to one joule per second. Here, the power is the rate of doing work that gives rise to the production of energy at a rate of one joule per second. Named after James Watt (1736-1819). 745.7 watts is equal to one horse power.

Wave cyclone

A cyclone which forms and moves along a front.

Wave-cycle

This is used to refer an extratropical cyclone. See 'cyclone'.

Wavelength

The shortest distance between adjacent crests or adjacent troughs of a train of waves. In case of a solitary wave, this is the length of the perceptible distortion. When waves travel in a medium, they do not necessarily proceed with velocity of light and velocity may vary with frequency. The distance between two successive similar points in a train of waves. The wavelength can be obtained when speed of the light is divided by its frequency. All waves travel with the velocity of light in space.

Waves

Regular or near regular disturbances of physical quantities such as temperature, electromagnetic field strength, etc., which repeat regularly in space and time.

Weather report (publication)

A periodically published information about the data recorded on meteorological observations for a location, place, region, etc., as per the IMD guidelines.

Weather vane (Wind wane)

An instrument that indicates the wind direction of 45 degrees or more, which takes place in less than 15 minutes (during thunderstorms).

Weather

- A day to day condition of the atmosphere at a given place and at a given instant of time in relation to temperature, humidity, wind velocity, and other meteorological parameters. Aspects involved are : 1. smaller regions like a village, city, etc., 2. smaller duration of time like a point of day or a complete day; and 3. expressed by numerical values of meteorological parameters. The other versions are as listed below.
- The state of the lower atmosphere at a given instant of time as defined by various meteorological elements.
- The behaviour of lower atmosphere with particular emphasis on those aspects of that behaviour which affect the lands and oceans and have a marked influence on the organisms that live upon the lands, with the waters of the Earth, or on the lower air.

Weathering

Physical disintegration and chemical decomposition of rocks and minerals on the earth's surface by natural processes like extreme temperature, wind, water and other biological agents etc.

Weed control efficiency

Also known as weed control index. It is the efficiency of the applied herbicide or a herbicidal treatment for comparison. Higher is the value higher is weed control efficiency. The formula $W.C.E = \dfrac{W_c - W_t}{W_c} \times 100$,

Where

W_c = Average weed weight or weed count in a weeded check plot.

W_t = Average wed weight or weed count in treated plot.

Weed control

Limiting weed infestations to minimum extent possible so that crops can be grown profitably or other operations can be conducted efficiently.

Weed eradication

Complete elimination of unwanted plants from the field. This is successful in small areas rather than larger areas.

Weed smothering efficiency (W.S.E)

It indicates smothering effect of inter crop on weed populations compared to solo cropping. It is given by the formula,

$W.S.E = \dfrac{Mdw - Idw}{Mdw} \times 100$

Where

Mdw = Average weed weight in solo cropping

Idw = Average weed weight in Inter cropping system

Weed

A plant growing where it is not desired or a plant for which economic value is not identified so far.

Weed index (W.I)

It is an index expressing the reduction in yield due to the presence of weeds in comparison with weed free situation. Lower the value higher the efficiency, The formula is, $W.I = \dfrac{YHW - Y_t}{YHW} \times 100$,

Where,

YHW = Average grain yield from hand weeded plot (weed free situation),

Y_t = Average grain yield from treated plot.

Weed management index (WMI)

An index that indicates how efficiently the weed is controlled in a given treatment compared to control and it is given by formula

$$WMI = \dfrac{\text{Percent yield over control}}{\text{Percent control of the weed}}$$

Weed persistance index (WPI)

The effect of weed control treatment is measured in terms of dry matter production in a crop or cropping system. Lower the value higher the efficiency and it is given by the formula

$$WPI = \dfrac{\text{Dry weight of weeds in treated plot}}{\text{Dry weight of weeds in control plot}}$$

Weight density

The force of attraction of the Earth on a given mass is the weight of that mass. So, the weight density is the weight per unit volume of any substance.

Westerlies

The wind systems flowing or moving predominantly from west to east.

Wet adiabatic lapse rate

An adiabatic change occurring in a wet air mass.

Wet bulb depression

The difference in degrees between the dry bulb temperature and the wet bulb temperature of a psychrometer.

Wet bulb temperature

The temperature registered by a wet bulb thermometer, of an air parcel would have, if cooled adiabatically to saturation at constant pressure. Here, the evaporation of water takes place because of the latent heat supplied by the air parcel.

Wet bulb

The bulb of the thermometer in a psychrometer which is covered with a moistened muslin.

Wet planting

A method of planting in which irrigation is given before planting and is usually followed in sandy soils.

Wet season

A period during which precipitation in excess of water requirement and water accumulated in the soil and in reservoirs.

Whirl wind

A relatively small rotating air mass occasionally embedded with dust and dried straw, etc.

White light

Light resolved into a continuous spectrum of wavelengths, so that all colours are in correct proportion to give whiteness.

White out

The condition of a complete whiteness arising from multiple scattering of sunlight between a low cloud layer and a snow covered surface. This whiteness resembles an incandescent light and under this situation the cloud and snow cover cannot be distinguished.

Wien's law

The law states that, the energy per unit wavelength is a function of absolute temperature and the wavelength. The wavelength at which the radiative energy flux density per unit wavelength is maximum turns out to be inversely proportional to the absolute temperature. Therefore, the maximum wavelength in terms of energy is equal to a constant (2827) divided by absolute temperature.

Wigwag

An instrument used to measure the solar radiation.

Wilt

Loss of freshness and drooping of foliage of a plant due to inadequate supply of moisture, excessive transpiration, or by a disease or nutrient deficiency which interferes with the utilization of water by the plant.

Wilting point

It is soil moisture condition at which the release of water to the plant root is just barely too small to counterbalance the transpiration losses. Of an oven-dry basis it is the moisture content of soil at which plants wilt and fail to recover their turgidity when placed in dark humid conditions. It is also called wilting coefficient or permanent wilting point. The availability of water is at −16 to −32 bars.

Wind break

A barrier in between the main crop and the moving wind. This barrier may be an another crop taller than the existing main crop or made up of dried stubbles, etc. The main objective of the wind break is to modify the microclimate of the main crop for better yields. It is to be noted that the wind breaks play a major role in all seasons by not only checking the evapotranspiration but also by providing the crop optimal growth conditions. Also see 'Shelter belts'.

Wind chill chart

A chart which indicates the effect of wind on a body. Also see 'Wind chill'.

Wind chill

This is a situation in which the temperature of a body is lower (when touched) than the measured value with a thermometer. This happens because of the cooling power of the wind with which it takes away the sensible heat and latent heat from that body.

Wind direction

The direction from which the wind is blowing at the place of observation. Expressed in points of the compass or in degrees measured clockwise from North.

Wind force

The force exerted by the air in horizontal motion on any object or construction.

Wind lull

Substantial decrease or reduction in wind speed for a short period.

Wind rose

A diagram resembling a flower indicating the relative frequencies of various wind directions for a given place or station and a period of time.

Wind shear

The stress applied to a body in the plane of any of its faces is 'shear'. Wind shear is the variation of any of the components of wind at a given location. Often this is referred to vertical shear of horizontal wind.

Wind shift

A change in wind direction of 45 degrees or more, in less than 15 minutes.

Wind speed

The rate of horizontal movement of air in relation to earth in the atmosphere expressed in metres per second, kilometers per hour, etc.

Wind stress

The force per unit area is stress. So, wind stress is the stress exerted on a surface by wind because of adjacent wind shear. Also see 'Wind shear'.

Wind vane

A device used to measure the wind direction.

Wind vector

A vector is any physical quantity that requires a direction to be stated in order to define it completely. So, a wind vector is an arrow representing wind velocity drawn to the point in the direction of the wind and with a length proportional to the wind speed.

Wind velocity

Velocity is the rate of motion in a given direction. So, wind velocity is a vector term to express the intensity of the wind, which includes both wind speed and wind direction.

Wind

The air in horizontal motion, relative to the surface of the Earth. The movement takes place from area of high pressure to areas of low pressure.

Windward

The direction from which the wind is blowing.

Winnowing

The process of separation of chaff and thrash from the mixture of grain and straw or bhusa. This can be done either with winnowing baskets or power operated winnowing fans.

Winter killing

Relatively high water transpiration rates in evergreens during a period when absorption of water can proceed only at a relatively slow rate may lead to a type of cold injury called winter killing.

Winter season

The period in which monsoon winds blow in winter months with continental origin.

Winter storm

Storm system that develop during the late fall early spring.

Winter

The period between the winter solstice and the vernal equinox.

Wire sonde

An atmospheric sounding instrument used to obtain temperature and humidity data.

Withering

The process of the physical and chemical disintegration and decomposition of rocks and minerals. On withering soils are formed.

Mortality of plants gradually by exhibiting wilting symptoms initially due to lack of soil moisture or attack from pests and diseases.

Working day
The day considered for the purpose of work is normally six to eight hours.

World data center
A center for collection and distribution of the geophysical data of the world with seven sub-centres, the main office of which is located at Washington, USA.

World Meteorological organization (WMO)
From weather prediction to air pollution research, climate change related activities, ozone layer depletion studies and tropical storm forecasting the world meteorological organization coordinates global scientific activity to allow increasingly prompt and accurate weather information and other services for public, private and commercial use, including international airline and shipping industries. Establishment by the United Nations in 1951, it is composed of 184 members.

 X

Xanthophyll

A yellow or orange carotenoid pigment associated with chlorophyll in chloroplats, also present in certain chromoplasts. The presence of such pigments said to be good sign as for nutritive value is concerned as they supply antioxidants.

Xerophyte

A plant with structural and physiological features which permit it to grow in a dry habitat, or under water stress conditions. Example Optunia Sp. Transfer of this xerophyte character among the varieties within a crop under rainfed conditions assumes significance in breeding programs.

Y

Year

The interval required for the earth to complete one revolution around the sun. (365 days, 6 hours,9 minutes, and 9.5 seconds). The calendar year begins at 12 o'clock midnight local time on the night of December 31st-January 1st. Currently, the Gregorian calendar of 365 days is used, with 366 days every four years, a leap year, is dependent on the seasons. It is the interval between two consecutive returns of the sun to the vernal equinox. In 1900, that took 365 days, 5 hours, 48 minutes, and 46 seconds, and it is decreasing at the rate of 0.53 second per century.

Yellow snow

Snow that is given golden, or yellow appearance by the presence of pollen of pine, cypress, etc.

Yield components (yield attributes)

Are components which finally make or control yield of any crop for example yield components of cereal grain yield are, individual grain weight grain, weight or test weight, grain number/ear, ear umber/plant and plant number/unit area.

Yield maximization

Agronomic practices adopted to get highest possible crop production per unit area per unit time without considering either the cost of production or net return. It is called as prize winning trail popularised among the resourceful farmers to get adapted recommended production technology.

Yield stability

The maintenance of yield at desired level over a period of time. Its index is measured by inverse of coefficient of variation, regression between yield and environmental index and disaster level estimated by probability of a system.

Yield potential

Full production capability of a crop under ideal conditions. Generally a variety expresses its potential yield under experimental station conditions under technical supervision where as at actual farm level it may not give its potential yield

Z

Zenith angle

The angle between a local line perpendicular to the surface, on which the radiation falls and the position of the Sun in the sky, from which the incident radiation falls.

Zenith

The point on the celestial sphere directly overhead.

Zephr

A pleasant, gentle and soft breeze.

Zero tillage (no-till)

The extreme form of soil and moisture conservation tillage is no-till (zero tillage) where in the new crop is planted in the residue of the previous crop without any prior tillage or seed bed preparation and is usually possible when all weeds, ratoons from stubbles are controlled by the use of herbicides.

Zeroplane displacement

This is the height (usually measured in metres) at which the turbulent exchange is assumed to start in an uniform crop canopy or tall vegetation. This is an aerodynamic term determined empirically and introduced to logarithmic wind profile to present the mean wind velocity profiles. This can also be applied to very rough surfaces and to represent the displacement of a profile above the crop canopies.

Zinc deficiency

Symptoms appear 2-4weeks after sowing as bleaching of the midrib of the emerging leaf, especially at base. Brown spots appear on the older leaves, which later on coalesce to give brown color. Tillering and growth are depressed and stunted. In severe form the plant dies. Zinc deficiency is associated with calcareus, alkali, peat, and volcanic soils and soils that are wet or water logged most of the year. Incidence is more severe where high rates of nitrogen and phosphorus are applied.

Zodiac

The position of the sun during the course of the year as it appears to move through successive constellations.

Zonal flow

The flow of air along a latitudinal component of existing, flow (west to east).

Zonal index

- In middle latitude a large scale flow particularly in troposphere.
- The measure of the strength of the westerly winds of the middle latitudes. Expressed as the horizontal pressure difference between 35 degrees and 55 degrees latitude.

Zonal

Parallel or along a line of latitude circle, i.e., West-East.

Zone of aeration

The layer of the Earth above the water table, containing air filled spaces through which water seeps. In geological studies, this layer is helpful in finding the occurrence of different events.

Zone of saturation

The layer beneath the zone of aeration in which all openings are filled with ground water. The water table is the top of the zone of saturation. Gravitational water form the major component in this water, in addition to seepage water.

Zulu time

One of several names for the twenty-four hour time which is used throughout the scientific and military communities.

<u>Units</u>

TABLE 1 : CONVERSION FACTORS

Non SI units	Multiply BY	Acceptable units To obtain
Acre	4.05×10^3	Square meter
Atmosphere	0.101	Mega pascal
cubic feet	0.028	cubic meter
cubic inch	1.64×10^{-5}	Cubic meter
Curie	3.7×10^{10}	Becquerel
degrees (angle)	1.75×10^{-2}	Radian
dyne	10^{-5}	Newton
erg	10^{-7}	Joule
foot	0.305	Meter
foot-pound	1.36	Joule
gallon	3.78	Liter
gallon per acre	9.35	Liter per hectare
gauss	10^{-4}	Tesla
gram per cubic centimeter	1.00	Megagram per cubic meter
gram per square decimeter hour (transpiration)	27.8	Milligram per square meter second
inch	25.4	Millimeter
kilogram per hectare (grain yield and forage)	0.001	Gram per square meter

Contd....

Non SI units	Multiply BY	Acceptable units To obtain
micro mole (H_2O) per square centimeter second (transpiration)	180	mg m^{-2} s^{-1}
micron	1.00	micrometer
mile	1.609	kilometer
mile per hour	0.447	kilometer
milligram per square decimeter hour (apparent photosynthesis)	0.0278	milligram per square meter second
milligram per square centimeter per second (transpiration)	10^4	milligram per square meter per second
millimho per centimeter	1.0	decisiemen per meter
ounce	28.4	gram
pound	454	gram
pound per acre	1.12	kilogram per hectare
quintal (metric)	10^2	kilogram
square centimeter per gram	0.1	square meter per kilogram
square feet	9.29 × 10^{-2}	square meter
square inch	645	square millimeter
square mile	2.59	square kilometer
square millimeter per gram	10^{-3}	square meter per kilogram
temperature (^{0}F −32)	0.556	temperature ^{0}C
temperature (^{0}C + 273)	1	temperature ^{0}K
ton (2000lb)	907	kilogram
ton (2000lb) per acre	2.24	megagram per hectare
centimol per kilogram, cmolkg^{-1} (ion exchange capacity)	1.0	milliequivalents per 100 grams
gram per kilogram	0.1	percent
megagram per cubic, Mg m^{-3} meter	1.0	gram per cubic centimeter
milligram per kilogram mg kg^{-1}	1.0	parts per million
P	2.29	P_2O_5
K	1.20	K_2O
Ca	1.39	CaO
Mg	1.66	MgO

TABLE 2 : EXAMPLES OF UNITS FOR GENERAL USE IN
PLANT SCIENCES

S.No.	Quantity	Application	Unit
1	Length	Soil depth, plant height	meter
2	Area	Land area	Square meter Hectare
		Leaf area	Square meter
		Specific surface area of the soil	Square meter per kilogram
3	Volume	Field	Liter, Cubic meter
4	Density	Soil bulk density	Mega gram per cubic meter
5	Electrical conductivity	Salt tolerance	Siemens per meter
6	Elongation rate	Plant	Millimeter per second Meter per day
7	Ethylene production	Nitrogen fixing activity	Nanomole per plant second
8	Extractable ions soil	Soil	Milligram per kilogram
9	Fertilizer rates soil	Soil	Grams per square Meter Kilogram per hectare
10	Fiber strength	Cotton fibers	Kilo Newton meter per kilogram
11	Flux density	Heat flow	Watts per square meter
		Gas diffusion	Mole per square meter second, gram per square meter second
		Water flow	Kilogram per square meter
12	Gas diffusivity	Gas diffusion	Square meter per second
13	Grain test weight	Grain	Kilogram per cubic meter
14	Hydraulic conductivity	Water flow	Kilogram second per cubic meter Cubic meter second per kilogram

Contd....

S.No.	Quantity	Application	Unit
15	Ion transport	Ion uptake	Mole per kilogram (of dry plant tissue) second Mole of charge per kilogram (of dry plant tissue) second
16	Leaf area ratio	Plant	squaremeter per kilogram
17	Length	Soil depth	Meter
18	Magnetic flux density	Electron spin resonance (ESR)	Tesla
19	Nutrient concentration	Plant	Milli mole per kilogram.
20	Photosynthetic rate	CO_2 amount of substance flux density	Micromole per square meter second (P)
21	Plant growth rate	------	Gram per square meter day
22	Resistance	Stomatal	Second per meter
23	Soil texture composition	Soil	Gram per kilogram Percent
24	Specific heat	Heat Storage	Joule per kilogram Kelvin
25	Thermal conductivity	Heat flow	Watt per meter Kelvin
26	Transpiration rate	H_2O flux density	Gram square meter second Cubic meter per square second
27	Volume	Field or laboratory	Cubic meter, liter
28	Water content	Plant Soil	Gram water per kilogram wet or dry tissue Kilogram water per kilogram dry soil. Cubic meter soil

Contd....

S.No.	Quantity	Application	Unit
29	X-ray diffraction patterns	Soil	Radians
30	Yield	Grain or forage yield	Gram per square meter Kilogram per hectare Mega gram per hectare Tonne per hectare Mass of plant or plant part